International Conference on Computational Information Technologies for Environmental Sciences (CITES-2017)

AA006694

IOP Conference Series: Earth and Environmental Science
Volume 96

Zvenigorod, Russia
4 - 7 September 2017

Editor:

Evgeny Gordov

ISBN: 978-1-5108-5240-2
ISSN: 1755-1307

Printed from e-media with permission by:

Curran Associates, Inc.
57 Morehouse Lane
Red Hook, NY 12571

proceedings.com

Some format issues inherent in the e-media version may also appear in this print version.

Copyright© (2017) by the Institute of Physics
All rights reserved. The material featured in this book is subject to
IOP copyright protection, unless otherwise indicated.

Printed by Curran Associates, Inc. (2017)

For permission requests, please contact the Institute of Physics
at the address below.

Institute of Physics
Dirac House, Temple Back
Bristol BS1 6BE UK

Phone: 44 1 17 929 7481
Fax: 44 1 17 920 0979

techtracking@iop.org

Additional copies of this publication are available from:

Curran Associates, Inc.
57 Morehouse Lane
Red Hook, NY 12571 USA
Phone: 845-758-0400
Fax: 845-758-2633
Email: curran@proceedings.com
Web: www.proceedings.com

Table of contents

Volume 96

International Conference on Computational Information Technologies for Environmental Sciences (CITES-2017)

4–7 September 2017, Zvenigorod, Russian Federation

Accepted papers received: 16 November 2017
Published online: 27 November 2017

Preface

Papers

Preface

These Conference Proceedings contain the selected papers of the 8th International Conference on Computational Information Technologies for Environmental Sciences (CITES) held on September 4-7, 2017 in Zvenigorod, Russia. The conference was preceded by Young Scientists School which was held on August 28 – September 2, 2017 in Tarusa. The main topic of the school was "Mathematical models of weather and climate forecasting". At the school 8 general and 9 special lectures were delivered by leading Russian and foreign specialists. In addition, school participants had 14 hours of practical training in computer classes, where they completed tasks within the preset 2 themes.

The conference was devoted to the state-of-the-art of topical modern modeling tools and information technologies developed for monitoring and assessment of regional environment variations including those caused by the Global Climate Change. Special attention was paid to Northern Eurasia at the targeted NEFI Workshop. During the conference, a total of 50 papers were presented and discussed in a series of successive presentations. All materials are presented in the conference proceedings and posted on the conference web-site http://www.scert.ru/en/conferences/cites2017.

All contributions, selected for publication in this volume, have been peer reviewed according to rigorous international standards. We are grateful to the reviewers for maintaining the standards and quality of the manuscripts throughout the reviewing process. However, the authors take full responsibility for the content of their papers.

The conference was organized by Institute of Monitoring of Climatic and Ecological Systems SB RAS, Institute of Numerical Mathematics RAS, Hydrometeorological Center of Russian Federation, A.M. Obukhov Institute of Atmospheric Physics RAS. It was supported by Russian Foundation for Basic Research, Federal Agency of Scientific Organizations, World Meteorological Organization and P.P. Shirshov Institute of Oceanology RAS under Megagrant of the Ministry of Education and Science of the Russian Federation «Extra-tropical hydrological cycle in the current and future climate: uncertainties and predictability». We gratefully acknowledge all our sponsors.

Editors

Professor Evgeny Gordov

RAS Corresponding Member Vasily Lykosov

Content from this work may be used under the terms of the Creative Commons Attribution 3.0 licence. Any further distribution of this work must maintain attribution to the author(s) and the title of the work, journal citation and DOI.
Published under licence by IOP Publishing Ltd

Peer review statement

All papers published in this volume of *IOP Conference Series: Earth and Environmental Science* have been peer reviewed through processes administered by the proceedings Editors. Reviews were conducted by expert referees to the professional and scientific standards expected of a proceedings journal published by IOP Publishing.

Content from this work may be used under the terms of the Creative Commons Attribution 3.0 licence. Any further distribution of this work must maintain attribution to the author(s) and the title of the work, journal citation and DOI.
Published under licence by IOP Publishing Ltd

International Conferences and Young Scientists Schools on Computational Information Technologies for Environmental Sciences (CITES) as a professional growth instrument

E P Gordov[1,2], V N Lykosov[3,4], E Yu Genina[1] and Yu E Gordova[1]

[1] Institute of Monitoring of Climatic and Ecological Systems SB RAS, 10/3 Akademichesky pr., Tomsk, 634055, Russia
[2] National Research Tomsk State University, 36, Lenin ave., 634050, Tomsk, Russia
[3] Institute of Numerical Mathematics RAS, 8, Gubkina str., 119333, Moscow, Russia
[4] Lomonosov Moscow State University, Research Computing Center, 1, bld.4, Leninskie Gory, 119234, GSP-1, Moscow, Russia

e-mail: gordov@scert.ru

Abstract. The paper describes a regular events CITES consisting of young scientists school and international conference as a tool for training and professional growth. The events address the most pressing issues of application of information-computational technologies in environmental sciences and young scientists' training, diminishing a gap between university graduates' skill and concurrent challenges. The viability of the approach to the CITES organization is proved by the fact that single event organized in 2001 turned into a series, quite a few young participants successfully defended their PhD thesis and a number of researchers became Doctors of Science during these years. Young researchers from Russia and foreign countries show undiminishing interest to these events.

1. Introduction

At present, environmental science is rapidly developing. Their conceptual and mathematical apparatus are expanded, models are developing, investigations have become complex and multidisciplinary. They are transforming into exact sciences and this process is proceeding in three rapidly developing areas: 1) the development of new tools for instrumental observations and data storage technologies; 2) creation of mathematical models describing Earth as a whole system and numerical analysis of its dynamics; 3) organization of all available information about the environment (large arrays of observational data and the results of numerical experiments, models and algorithms, and ways to display this information) in information-computational and information systems. The MODAS conference held in Irkutsk in 2001 was one of the first conferences dedicated to the problems facing the environmental sciences. One of the most important results of this conference was the realization of the necessity to train young scientists who can meet the above challenges of the contemporary environmental sciences and work in multidisciplinary teams.

The main driver to elaborated new approaches was the fact that university graduates have no skill in working with contemporary climate models, large data archives and computational resources. The concept has been developed and implemented that uses the "two-component" approach to training young scientists in the field of environmental sciences (http://www.scert.ru/en/conferences/cites/). Thanks to suggestion of academician V.P. Dymnikov, the first component is an event consisting of a thematic school for early carrier students with courses of general lectures on 2-3 selected specific topics in the field of information-computational technology and the implementation of practical

Content from this work may be used under the terms of the Creative Commons Attribution 3.0 licence. Any further distribution of this work must maintain attribution to the author(s) and the title of the work, journal citation and DOI.
Published under licence by IOP Publishing Ltd

training for one of the courses, then - a scientific conference in which highly-qualified specialists, including invited lecturers, take part together with young scientists. The second component is a multidisciplinary conference comprising traditional elements of early carrier scientist's school [1].

Experience in the implementation of the developed concept [2] has shown that young scientists who have completed such an educational "cycle" acquire the experience necessary to meet the challenges of modern environmental sciences, and are quickly included in multidisciplinary studies carried out by national and international teams. We believe that this experience can be useful and applicable in other topical areas.

2. Approach

The CITES scientific event is held every two years in different cities of Russia, mainly in its Siberian part. It gives an opportunity to those students in regions who have no travel grants to receive first-hand knowledge and practice. Its first part is a thematic school for early carrier students with courses of general lectures on selected specific topics and practical training for one of the courses. The second component is a conference in which highly-qualified specialists take part together with young scientists.

Every school starts with a day of deep immersion in a school subject from the selected area of informational computational technologies. It is called an "Academicians day", during which several members of the Russian Academy of Sciences deliver hour-and-a-half lectures on key "hot" topics of the domain. During following days invited lecturers deliver their lecture courses to school participants and one of these courses is supported by a specially designed training, which consists of a set of small research projects devoted to topical environmental science problems. Solving training tasks requires competent usage of selected modern computation and information technologies. All calculations needed to complete the training tasks are performed on supercomputers of a network of Russian computational centers, e.g. in Novosibirsk, Tomsk and Moscow. In this way young scientists have an opportunity to use modern high-performance computing systems that are necessary at present to process large datasets.

To perform their practical tasks, school participants are divided into small groups of 3-4 persons. The persons in such a group are usually from different fields of science, e.g. meteorology, geography, computer sciences, mathematics, and often with a different level of expertise. It gives them an experience of a team work and cooperation that is an important skill at present, when scientific projects are implemented by large interdisciplinary consortia. During this team work personal and professional links are established which may be useful for school participants in future. After performing their small research project each group prepares a short presentation of results obtained and presents them during the subsequent conference. It gives them an experience in public reporting of their results. In addition to the school reports the students of the school present reports on their main work. All students who attended lectures and completed practical assignments receive personal certificates.

The conference CITES is usually devoted to the-state-of-the-art of topical modern modeling tools and information technologies developed for Earth System Sciences applications. The Conference is run as a set of consequent sessions with invited lectures, invited oral and contributed poster paper presentations so participants have an opportunity to attend any report without overlapping sessions. Some of these conferences included special workshops on current international research projects to give young scientists an impression of new scientific frontiers. Such workshops were devoted to following topics: Siberia Integrated Regional Study development, Atmospheric Composition and Air Quality, Man-made Environmental Risks in Siberia, Northern Eurasia Earth Science Partnership Initiative, etc.

Even the first CITES event set the bar high and the students rose to meet it. School participants are selected on the basis of specially developed questionnaires containing the following items: education, publications, research interest areas and supervising scientist's recommendations. The level of a proposed conference report is also important.

Due to its high standards, the CITES event was highly appreciated by national and international science support organizations. It resulted in several grants of INTAS (The International Association for the Promotion of Co-operation with Scientists from the New Independent States of the Former Soviet Union) and RFBR (Russian Foundation for Basic Research). It also was put on the INTAS list of permanent supported young scientist school just before its closure.

3. Results

CITES events trace their origin to 2003. Subjects of the school are determined each time by the most pressing issues of environmental sciences.

Thus, CITES-2003 (held in Tomsk) was devoted to the state and basic fundamental and applied problems of climate modeling and its changes, CITES-2005 (Novosibirsk) paid special attention to methods of data assimilation and methods of remote sensing of the environment, CITES-2007 (Tomsk) – to transport of impurities in the atmosphere, CITES-2009 (Krasnoyarsk) was devoted to statistical methods in the environmental tasks**,** CITES-2011 (Tomsk) – to monitoring and diagnosis of modern climate changes on regional scale, CITES-2013 (Petrozavodsk) focused on the study of the ocean and climate interrelations, the main topic of CITES-2015 (Tomsk) was aerosols, small gas impurities and climate and CITES-2017 (Tarusa/Zvenigorod) considered mathematical models of weather and climate forecasting. Practical courses correspond to general school topics. At the first event (CITES-2003) the students were acquainted with accessible information resources, statement of computing tasks, working on them. Then practical courses were the following: data assimilation and environmental processes modeling (CITES-2005); numerical methods of solving transport equation, modeling of chemically interacting atmospheric gases (CITES-2007); statistical methods in problems of atmospheric and ocean physics, data assimilation: dynamic-stochastic approach [3] and Kalman filter [4] (CITES-2009); modeling of atmospheric mesoscale processes on the basis of modern parallel computing technologies, numerical methods for calculating regional impurity transport (CITES-2011); long-term variability in the climate system (CITES-2013); chemical data assimilation [5] (CITES-2015); investigation of the influence of various parameters in the global SLAV model on a long-term weather forecast [6, 7], climate modeling with the INM RAS climatic model [8, 9] (CITES-2017).

It should be noted that the lectures delivered at the CITES schools, as well as practical courses specially developed for these schools, laid the basis for courses that currently are included in the curricula of the Lomonosov Moscow State University, the Moscow Institute of Physics and Technology and Tomsk State University. In addition, the lectures on the interaction of the atmosphere and the hydrosphere delivered during the CITES-2013 held in Petrozavodsk became the basis of a collective monograph [10]. Lectures are presented in the form of separate chapters, which consider the following topics: the role of the ocean in the observed and possible in the future climate variability; models and methods used in large-scale atmospheric-ocean interactions; the problem of assimilation of observational data in models of ocean and sea circulation; short-period response of the upper layer of the ocean to the atmospheric effect; sensitivity of climatic models to small perturbations; regional and local aspects of the interaction of the atmosphere and the hydrosphere (lakes and climate, wind waves on the oceans and seas, hydrological processes on land); computational-information technologies for the analysis of climatic changes and game-theoretic models of cooperation in the management of marine bioresources. The final chapter of the monograph presents the results obtained by the school participants during practical courses on climate change modeling. This practical course has been prepared and implemented by Doctor of Sciences in Physics and Mathematics E.M. Volodin, a leading scientist of the Institute of Numerical Mathematics of the Russian Academy of Sciences, with a team of 12 young school participants. They considered the following tasks: 1) determination of the equilibrium sensitivity of the climate model to quadrupling the concentration of CO_2; 2) modeling climate changes when large amounts of fresh water enter the North Atlantic; 3) modeling climate change due to a change in the inclination of the Earth's rotation axis to the ecliptic plane; 4) modeling effect of compensation for global warming by introducing into the stratosphere sulfur-containing substances. Reports prepared by school participants and presented at the conference made this final chapter.

It became a good tradition of the CITES events to organize a competition of young scientist's reports. The Program Committee members assess presentations and posters and choose three winners. These persons granted an opportunity to publish their papers in high-ranked journals. For example, winners of CITES-2015 and CITES-2017 events were invited to publish papers in Russian Meteorology and Hydrology journal.

One on the main goals of CITES events is training young scientists and contribute to their professional skill growth. Within CITES history, quite a few young participants (more than 25 persons), who were students and post-graduates, successfully defended thereafter their PhD theses. Now they participate in the events as lecturers and training instructors. A number of researchers (7 persons) became Doctors of Science. Chosen approach to forming groups is also proved its success, which is confirmed by several research projects realized by the consortia made on the basis of such school groups.

This biannual CITES event, consisting of Early Career Scientists School and scientific conference, became well-known not only in Russia but also abroad. Scientists from such leading foreign centers as US NOAA, US Environmental Protection Agency (EPA), Massachusetts Institute of Technology, University of New Hampshire, National Institute for Environmental Studies (Japan), Potsdam Institute for Climate Impact Research (Germany), Danish Meteorological Institute (Denmark)) regularly participate in CITES conferences, where they deliver lectures and present studies carried out at these organizations. Russian leading organizations also always have participating in organization of the CITES events. First of all, Institute of Numerical Mathematics RAS, Research Computing Center of the Lomonosov Moscow State University, Institute of Monitoring of Climatic and Ecological Systems SB RAS are permanent organizers of the event. Among other regular participants there are scientists form Hydrometeorological Center of Russian Federation, P.P Shirshov Institute of Oceanology RAS, A.M. Obukhov Institute of Atmospheric Physics RAS, Institute of Computational Mathematics and Mathematical Geophysics SB RAS, Siberian Federal University, V.E. Zuev Institute of Atmospheric Optics, Tomsk State University and other organizations. Computing resources of high-performance computers of Tomsk State University, Institute of Computational Mathematics and Mathematical Geophysics, Lomonosov Moscow State University and Institute of Numerical Mathematics RAS have been used for practical courses.

About 1070 researchers took part in CITES conferences, among them there were 95 foreigners. Half of the CITES conferences participants have been young scientists.

From the very beginning CITES event had been supported by various funds. Thus, CITES-2003 and CITES-2005 have been financed by EC grants. CITES-2007 event was supported within the APN Project "Capacity Building to Study Interrelations between Atmospheric Composition, Anthropogenic Load and Climate Change in Northern Asia" (http://project.enviromis.scert.ru/apn/). This project encompassed regional activity oriented to the Northern Asia territory and had been fulfilled by institutions of four countries: Russia, Japan, Kazakhstan, and Uzbekistan. It enhanced involvement of regional young scientists into this important study and assisted to their further career development. The Project contributed to the regional capacity building activities and to sustainable development in the region by its assertive data and scientific results dissemination activity. A number of young scientists from APN target countries had been financially supported to participate in this scientific-educational event. The APN Project «Human Impact on Land-Cover Changes in the Heart of Asia» (http://project.enviromis.scert.ru/asia/) was fulfilled by organizations from Russia, USA, and Mongolia. This project was focused only on four (though quite typical) regions in West Siberia and Mongolia in the light of investigation of human impact on land-cover change. Within the Project, land-cover change and disturbance indices maps were developed for the study regions. Investigations were carried out using satellite image analysis and field observations. Project results were disseminated at several international conferences, including CITES-2009, ENVIROMIS-2010,

European Geoscience Union General Assembly 2010 and American Geophysical Union Fall Meeting 2010. Participation of the project team members in these events was financially supported.

In order to disseminate research results among researchers and public, we publish CITES conferences proceedings. Electronic versions of these proceedings are posted on the conference web-site (http://www.scert.ru/en/conferences/cites/) and selected papers are published in special issues of journal "Computational Technologies", 2008 [11]. Presentations of lectures delivered at schools and reports made at conferences are also posted on the relevant conference web-site (http://www.scert.ru/ru/conference/).

4. Conclusion

The described above approach has shown its viability and proved that the CITES events comprising International Conference and Young Scientists School on Computational Information Technologies for Environmental Sciences are contributing a process of young scientists professional skill growth. This year, CITES-2017 conference was the 8th in the series and some delivered papers are presented in this Proceedings volume.

It should be added that at different periods of time, these conferences were supported by INTAS, European Commission, WMO, and the Asia-Pacific Network for Global Change. They are also regularly supported by the Russian Foundation for Basic Research. Despite the fact that this financial support is modest, researchers show undiminishing interest to these events. This allows us to hope that the organizers are on the right way.

Acknowledgment

Organization of CITES-2017 additionally to the SB RAS fundamental research program 138.2 was supported by RFBR grant (No.17-35-10225), as well as targeted funding from WMO and Federal Agency of Research Organizations.

References

[1] Gordov E P, Lykosov V N, Gordova Yu E and Genina E Yu 2016 Multidisciplinary ENVIROMIS conferences as a tool connecting climate information producers and users *IOP Conf. Series: Earth and Environmental Science* **48** 012001 doi:10.1088/1755-1315/48/1/012001

[2] Gordov E P, Kabanov M V and Lykossov V N 2006 Information-computational technologies for environmental science: young scientists training *Computational technologies* **11** (Special issue) 3–15

[3] Bogoslovskii N N, Tolstykh M A and Shlyaeva A V 2008 Data assimilation for surface and soil variables in the global semi-Lagrangian NWP model *Computational Technologies* **13** (Special issue 3) 111–16

[4] Klimova E G 2011 The use of ensemble Kalman filter for the planning of adaptive observations *Russian Meteorology and Hydrology* **36** (5) 11–19

[5] Penenko A V 2015 Data assimilation in transport and transformation models of atmospheric chemistry *Proc. Int. Conf. on Urgent Problems of Computational and Applied Mathematics* (Novosibirsk: Institute of Computational Mathematics and Mathematical Geophysics SB RAS) pp 555–61

[6] Shashkin V V, Tolstykh M A, Ivanova A R and Skriptunova E N 2017 SL-AV Atmospheric Model Version Using –p Hybrid Vertical Coordinates *Russian Meteorology and Hydrology* (9) 24–35

[7] Tolstykh M A *et al* 2015 Development of the multiscale version of the sl-av global atmosphere model *Russian Meteorology and Hydrology* **40** (6) 374–82

[8] Volodin E M *et al* 2010 Earth system model of INM RAS *Russ. J. Numer. Anal. Math. Modelling* **25** (5) 531–45

[9] Volodin E M, Diansky N A and Gusev A V 2013 Simulation and prediction of climate changes in the 19th to 21st centuries with the Institute of Numerical Mathematics, Russian Academy of Sciences, model of the Earth climate system *Izvestiya. Atmospheric and Oceanic Physics* **49** (4) 347–66
[10] *Models and Methods for Atmosphere – Hydrosphere interaction* 2014 ed V P Dymnikov, V N Lykosov and E P Gordov (Tomsk: Tomsk State University Publishing House) 376 p
[11] Gordov E P and Lykossov V N 2008 ICT for environmental sciences: synthesis of science and education *Computational Technologies* 13 (Special issue 3) 3–11

Inversion effects on wind and surface pressure in atmospheric front propagation simulation with a hyperbolic model

M S Yudin

Institute of Computational Mathematics and Mathematical Geophysics SB RAS, Novosibirsk, Russia

e-mail m.yudin@ommgp.sscc.ru

Abstract. In this paper the effects of an inversion layer in a stratified atmosphere on the surface wind speed and pressure are investigated with models based on the compressible Navier-Stokes equations in two dimensions. Artificial compressibility is introduced into the models in order to make the governing equations hyperbolic. For comparison with available simulation data, the physical processes under study are assumed to be adiabatic. Plain orography is considered in surface pressure simulations with a finite-difference version of the model, while surface wind speed effects are estimated in artificial cold front propagation over a hill with a finite-element version of the model. The front surface is described in both models by an equation for advection of a scalar substance, which is solved with a third-order semi-Lagrangian procedure. The results of simulations show various meteorological effects in agreement with observations and in accordance with a theory proposed by Charba [3].

1. Introduction

A front in the atmosphere is a phenomenon of gravitational flows that take place in a variety of forms: breeze fronts, storm flows etc [1,8]. Phenomena of great importance are cold atmospheric fronts propagating near the surface with high speeds. These fronts may be retarded and changed in shape under the influence of the underlying surface and stratification of the atmosphere. The retardation of a frontal system on the windward side of a mountain is a commonly observed phenomenon [11, 12].

Atmospheric gravity currents occupy a wide range of length scales from several meters to thousands of kilometers. These currents man be subdivided into classes varying from micro- to macro-scales. Meso-scale flows lie in an interval approximately from two to two thousand kilometers [9]. The flows of interest in the present study are meso-scale currents. These flows are relatively shallow: they belong to the atmospheric boundary layer and range only a few kilometers from the surface in the lower atmosphere.

Stratification and inversion effects on surface pressure in the propagation of an atmospheric gravity current (cold front) over flat terrain are estimated in the present study with a non-hydrostatic finite-difference and finite-element models of atmospheric dynamics. Artificial compressibility is introduced into the models in order to make the equations hyperbolic.

The property of hyperbolicity is important for the correct formulation of proper boundary conditions. It also allows efficient implementation of the computational algorithms and methods used in the simulation. An important step forward was recently made by Peshkov and Romenski [2] in formulating a hyperbolic model for viscous flows.

In this paper the effects of introducing an inversion layer into a stratified atmosphere on the surface wind speed and pressure are investigated with models based on the compressible Navier-Stokes equations in two dimensions.

Artificial compressibility is introduced into the models in order to make the model equations hyperbolic. For comparison with the available simulation data, the physical processes under study are

Content from this work may be used under the terms of the Creative Commons Attribution 3.0 licence. Any further distribution of this work must maintain attribution to the author(s) and the title of the work, journal citation and DOI.
Published under licence by IOP Publishing Ltd

IOP Conf. Series: Earth and Environmental Science **96** (2017) 012002 doi:10.1088/1755-1315/96/1/012002

assumed to be adiabatic. Plain orography is considered in surface pressure simulations with a finite-difference version of the model, while surface wind speed effects are estimated for artificial cold front propagation over a hill with a finite-element version of the model. The front surface is described in both models by an equation for advection of a scalar substance, which is solved with a third-order semi-Lagrangian procedure. The results of simulations show various meteorological effects in agreement with observations and confirm a theory proposed by Charba [3].

The results of simulations of surface pressure under stable stratification with an inversion layer are presented. The model is based on spatial discretizations that conserve some important quantities of the phenomena under study like momentum and scalars. Also, an efficient procedure is used to calculate the advection of scalars. Also, in the present paper, a preliminary investigation is carried out to simulate cold front propagation over a steep hill in two dimensions with a finite-element model.A 2D finite-element model based on triangular elements is used to simulate the of cold front propagation over an idealized orographic obstacle.

2. 2D model equations

The Navier-Stokes equations for a compressible air flow are used here for the calculation of gravity flows in a stratified atmosphere An exact formulation of the equations is given in [15].A more detailed description of the model can be found, for instance, in [17].In the present study, a two-dimensional finite-difference and finite-element versions of the model are employed [17].The time discretization is similar to that proposed in [13].It is also described in paper [16].

For the simulations a small-scale non-hydrostatic model is used, which is essentially a form of the Navier-Stokes equations in the Boussinesq approximation. A version of the equations relevant for a compressible fluid is used here. Since sound waves in the atmosphere have small effects on gravity flows like fronts, a compressible version of the Navier-Stokes equations is taken here to make the equations hyperbolic, and also for pure technical reasons. The equations are discretized with a splitting method in time, and the compressible form of the equations is suitable for an efficient splitting. Sound waves are then eliminated by the use of an Asselin-type filter [4]. The time discretization of the governing equations may be written as follows:

$$\delta_t U + \frac{\partial}{\partial x}\overline{P}^t + \frac{\partial}{\partial \xi}(G^{13}\overline{P}^t) = -ADVU,$$

$$\delta_t W + \frac{1}{G^{1/2}} + \frac{\partial \overline{P}t}{\partial \xi} + \frac{g\overline{P}^t}{Cs^2} = BUOY - ADVW,$$

$$\frac{1}{Cs^2}\delta_t P + \frac{\partial}{\partial x}\overline{U}^t + \frac{1}{G^{1/2}}\frac{\partial \overline{W}^t}{\partial \xi}$$

$$+\frac{\partial}{\partial \xi}(G^{13}\overline{U}^t) = PFT,$$

$$\overline{f}^t = \frac{1+\alpha}{2}f^{t+\Delta t} + \frac{1-\alpha}{2}f^{t-\Delta t},$$

$$\delta_t f = \frac{f^{t+\Delta t} - f^{t-\Delta t}}{2\Delta t}$$

The problem examined in this study has very small physical diffusion, and the governing equations can be considered as almost inviscid and adiabatic. Special operators of space discretization are used to provide conservation of momentum and scalars in the finite-difference model (for a detailed description see [15,17]. Triangular elements are used in the finite-element model [16]. In both model versions a suitable scheme for the advection of a scalar, e.g. temperature, is implemented.

For fully hyperbolic problems explicit schemes have been widely used in the past and will probably continue to be used in the future. Traditionally, problems for which the governing equations are hyperbolic have traditionally been solved by methods of characteristics. Sometimes the conventional methods of characteristics are slower than competitive finite difference methods. However, some finite difference schemes use a form of the governing equations such that knowledge of the characteristic locations can be exploited in an efficient way [10].

In this paper, the advection of a scalar like temperature is treated with an efficient semi-Lagrangian finite difference scheme of third-order [10] that is used as a reasonable compromise between cost and accuracy. A detailed description of the scheme can be found in [4].

3. FDM version : smooth surface pressure simulation

The results of a series of calculations are presented here to simulate the stratification and inversion effects on the surface pressure in the propagation of an atmospheric gravity current (cold front) over flat terrain .The model parameters are taken from paper [11]. The calculation domain is 25x25 km. In contrast to [11], where an atmospheric front is generated by a volume of cold air, in the present study the front is initially given in the form of a step- function of 400 m in height.

Figure 1 shows the initial location of the front. Figure 2 shows the results of calculations of the surface pressure at the point of 12 km under neutral stratification. In Figures 3 and 4 the time evolution of the surface pressure is given at the same spatial point in an atmosphere under stable stratification and in an atmosphere under stable stratification with an inversion layer.

Under stable stratification the front moves faster and shows an abrupt pressure jump at the point of observation. The introduction of an inversion layer into the atmosphere increases the pressure further.In contrast to the slow evolution of surface pressure under neutral stratification, there is a considerable pressure jump in a stable atmosphere. This effect is increased by the introduction of an inversion layer. This phenomenon was explained by Charba [3].The results of calculations in the present paper are in good agreement with Charba's theory.

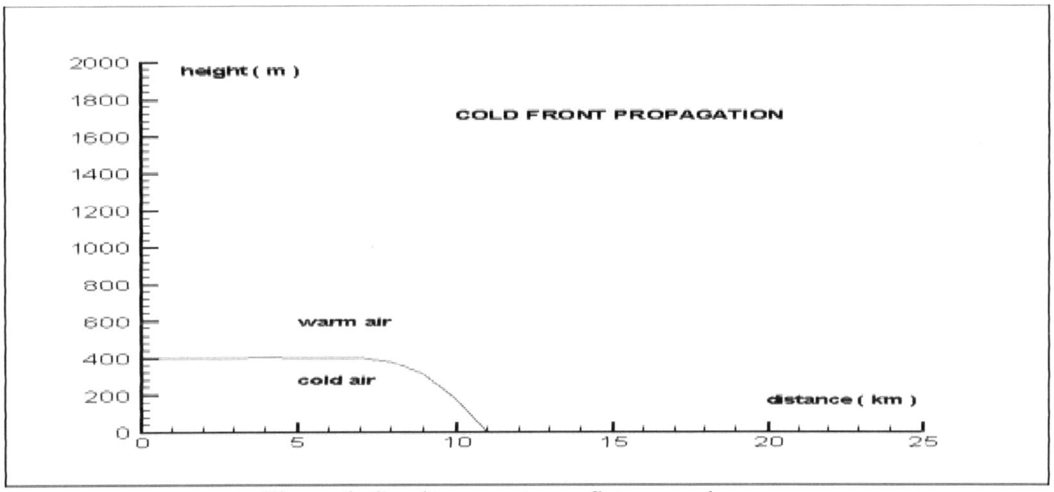

Figure 1. Gravity current over flat orography.

IOP Conf. Series: Earth and Environmental Science **96** (2017) 012002 doi:10.1088/1755-1315/96/1/012002

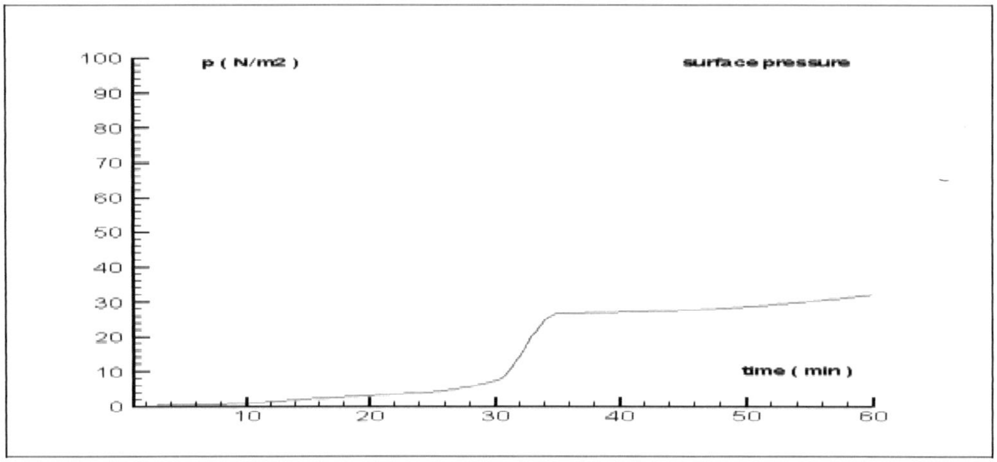

Figure 2. Surface pressure at 12 km. Neutral stratification.

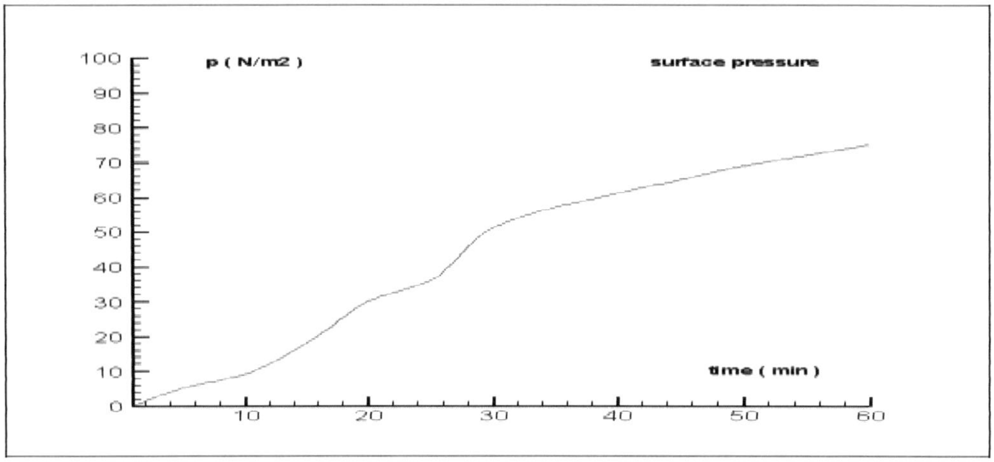

Figure 3. Surface pressure at 12 km. Stable stratification.

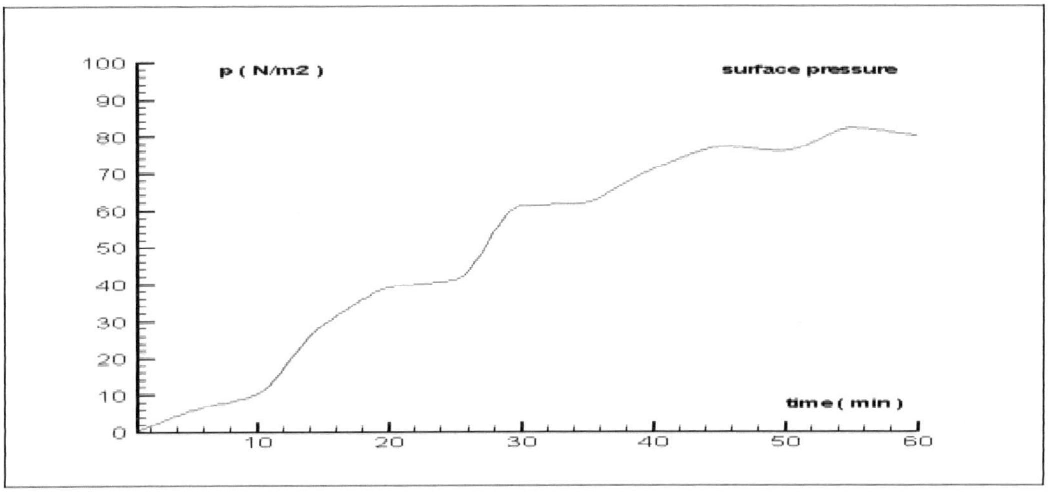

Figure 4. Surface pressure at 12 km. Stable stratification with inversion.

4. FEM version : steep surface wind speed simulation

In order to make the study more complete, in this section the results of paper [4] related to the effects of inversion on surface wind speeds in cold front motion over a steep obstacle are shown. Specifically, the effects of introducing an inversion layer into a stably stratified atmosphere are studied with a finite-element model described in paper [4]. The model parameters are taken as in paper [12]. In this paper, a cold front propagating over an axially-symmetric obstacle in the form of a bell-shaped hill is simulated with a finite-difference model. The height of the hill is 600 m. The calculation domain is 25x2 km. In contrast to [12], where the front is generated by a volume of cold air, in the present study the front is initially given in the form of a step-function of 400 m in height.

An inversion layer of 300 m in height is placed over the isolated hill described above. The atmosphere has the standard stable stratification of 3.5K/100 m, whereas the stratification of the inversion layer is 1K/100 m. The results of a series of calculations of cold front propagation over the hill and a plain are shown in Table 1. The calculated values of windward and leeward speeds are given in the last columns.

It can be seen from the table that the introduction of the inversion layer produces a significant decrease in the front speed both for the currents over the obstacle and those over flat orography. This is in agreement with the results of papers [11,12], where the front was generated by a volume of cold air. As in [11,12], the introduction of an elevated inversion layer causes a considerable intensification in the motion of vertical air and increases the meso-scale heat flux. This results in a more intensive entrainment.

Table 1. Calculated windward and leeward speeds of cold front propagation over a hill and a plain. Stable stratification.

obstacle height (m)	initial front height (m)	stratification (K / 100 m)	windward speed (m/s)	leeward speed (m/s)
600	400	no inversion	4.9	2.7
600	400	Inversion	4.4	2.2
0	400	no inversion	5.1	5.1
0	400	Inversion	4.6	4.6

5. Conclusions

The results of the simulations described above have been obtained with two types of models: a 2D non-hydrostatic finite-difference meteorological model and a 2D finite-element model. Both models are dicretized versions of a non-hydrostatic model formulated as the Navier-Stokes equations in the Boussinesq approximation written in a compressible form.

The 2D finite-element model is based on triangular elements. In this paper an application of the model was made to simulating cold front propagation over an idealized hill-type obstacle in a stratified atmosphere with an inversion layer over an isolated hill. The study was performed under stable stratification in and beyond the inversion layer. It has been shown that the introduction of the inversion produces a significant decrease in the front speed both for the currents over the obstacle and those over flat orography.

The 2D finite-difference model is based on spatial discretizations that conserve some important quantities of the phenomena under study, atmospheric gravity currents. Also, an efficient procedure is used in both versions of the model to calculate the advection of scalars.

The change in stratification from neutral to stable in the propagation of the cold atmospheric front has shown a time evolution of the surface pressure that is in good agreement with the available observational data [11]. Also, in contrast to the slow evolution of the surface pressure under neutral stratification, there is a considerable pressure jump in a stable atmosphere. This effect is increased by the introduction of an inversion layer.This phenomenon was explained by Charba [3].The results of calculations in the present paper are in good agreement with Charba's theory.

Acknowledgements

This work was supported by the Russian Foundation for Basic Research under grant 17-01-00137 and the Presidium of RAS under programs I.33P, II.2P/3-2 and II.2P/3-3.

References

[1] Schultz D M 2005 A Review of Cold Fronts with Prefrontal Troughs and Wind Shifts *Mon. Wea. Rev.* **133** 2449–72

[2] Peshkov I and Romenski E 2016 A hyperbolic model for viscous newtonian flows *Continuum Mechanics and Thermodynamics* **28** 85–104

[3] Charba J 1974 Application of a gravity current model to analysis of squall-line gust fronts *Mon. Wea. Rev.* **102** 140–56

[4] Yudin M S 2016 A numerical study of gravity waves in the atmosphere: smooth and steep orography effects *IOP Conference Series: Earth and Environmental Science* **48** DOI http://dx.doi.org/10.1088/1755-1315/48/1/012024

[5] Marchuk G I 1974 *Numerical Solution of Atmospheric and Oceanic Problems* (Leningrad: Gidgometeoizdat) p 303

[6] Marchuk G I 1982 *Mathematical Models in Environmental Problems* (Moscow: Nauka) p 319

[7] Penenko V V and Aloyan A E 1985 *Models and Methods for Envronmental Protection Problems* (Novosibirsk: Nauka) p 256

[8] Pielke R A 2002 *Mesoscale Meteorological Modeling* (San Diego: Academic Press) p 676

[9] Davies H C 1984 On the Orographic Retardation of a Cold Front *Beitr. Phys. Atmos.* **57** 409–18

[10] Ritchie H 1987 Semi-Lagrangian Advection on a Gaussian Grid *Mon. Wea. Rev.* **115** 136–46

[11] Bischoff-Gauss I and Gross G 1989 Numerical Studies on Cold Fronts. Part 1: Gravity Flows in a Neutral and Stratified Atmosphere *Meteorol. Atmos. Phys.* **40** 150–58

[12] Bischoff-Gauss I, Gross G and Wippermann F 1989 Numerical Studies on Cold Fronts. Part 2: Orographic Effects on Gravity Flows *Meteorol. Atmos. Phys.* **40** 159–69

[13] Ikawa M 1988 Comparison of Some Schemes for Non-Hydrostatic Models with Orography *J. Meteor. Soc. Japan* **66** 753–76

[14] Schumann U 1987 Influence of Mesoscale Orography on Idealized Cold Fronts *J. Atmos. Sci.* **44** 3423–41

[15] Yudin M S 2011 Propagation of a Gravity Current in the Atmosphere over a Valley *Bull. Novos. Comput. Center* **14** 65-70

[16] Yudin M S and Wilderotter K 2006 Simulating Atmospheric Flows in the Vicinity of a Water Basin *Computational Technologies* **11** 128–34

[17] Yudin M S 2012 Comparison of FDM and FEM Models for a 2D Gravity Current in the Atmosphere over a Valley *Bull. Novos. Comput. Center* **13** 95–101

Experimental study of heat and momentum exchange between a forest lake and the atmosphere in winter

K V Barskov[1,2], R V Chernyshev[2], V M Stepanenko[2], I A Repina[1,2], A Yu Artamonov[1], S P Guseva[2] and A V Gavrikov[3]

[1] A.M.Obukhov Institute of Atmospheric Physics RAS, Moscow, Russia.
[2] Lomonosov Moscow State University, Moscow, Russia
[3] P.P.Shirshov Institute of Oceanology RAS, Moscow, Russia

Barskovkv@gmail.com

Abstract. The article presents the results of an experimental study of turbulent heat exchange between the surface of a frozen lake surrounded by forest and the atmospheric boundary layer. Heat and momentum fluxes were measured at three levels by an eddy covariance (EC) technique. Additionally, the heat fluxes were estimated by a surface energy balance method using a temperature profile measured in the snow cover and net longwave and shortwave radiation. The results of the measurements show that the eddy covariance fluxes correlate well with those obtained by the surface energy balance method, with a tendency of underestimation. The presence of wind-shear effects at treetop height demonstrated recently in a Large Eddy Simulation (LES) [3] was supported in our measurements by the fact that the momentum flux increased with height from the surface. The negative sensible heat flux increased with height most of the time. We suggest that this phenomenon may partially be caused by the high negative heat fluxes above the surface formed when warm advection occurs at altitudes of ~100 m. During the warm advection events, Monin-Obukhov similarity theory (MOST) fails to reproduce the sharp increase of the negative heat flux at the surface layer. Beyond the warm advection events, the MOST calculations agree well with the EC fluxes, however, with some systematic underestimation bias.

1. Introduction

Lakes cover a considerable part of Northern Eurasia and North America and make a significant contribution to heat and gas exchange between the atmosphere and the surface. To parameterize these exchanges, especially with subgrid-scale surface inhomogeneities, such as small lakes bounded by steep topography, is important for improvement of weather and climate models [1]. Lakes smaller than 10 km^2 account for about 54% of the global area of inland waters [2]. Many lakes in Northern Russia and Europe are surrounded by forest, and standard approaches of calculating heat exchange in the atmospheric boundary layer leads to significant errors due to the inhomogeneity of the underlying surface. In particular, landscape inhomogeneity causes changes in the statistical moments of turbulence, compared to flow over homogeneous surfaces. Thus, the application of Monin-Obukhov similarity theory (MOST) looses its theoretical grounds. Extending MOST for heterogeneous surfaces is difficult due to the lack of experimental data for the wide range of possible wind shears, stratifications, and surface discontinuity patterns.

Experimental measurement of fluxes in complex landscapes is also difficult. In particular, the results of a Large Eddy Simulation (LES) show that at distances from the forest edge less than 15-20 treetop heights the local eddy covariance (EC) momentum fluxes above the surface are caused by the velocity shear in the windstream approximately at treetop height and may differ greatly from the surface frictional stresses [3]. Such effects are inherently connected to the long-standing problem of surface energy imbalance in the EC measurements. Specifically, in most EC field experiments the energy of net

Content from this work may be used under the terms of the Creative Commons Attribution 3.0 licence. Any further distribution of this work must maintain attribution to the author(s) and the title of the work, journal citation and DOI.
Published under licence by IOP Publishing Ltd

radiation minus the change of active layer heat content (ice, snow, water, or soil surface) is greater than the energy transported by turbulent fluxes of sensible and latent heat [4].

This paper presents the results of experimental studies of the turbulent structure of the atmospheric boundary layer over a small frozen lake surrounded by forest. The heat exchange of the surface with the atmosphere was estimated by various methods, specifically, eddy covariance, surface energy balance and MOST-based techniques. The experiment was conducted in wintertime over an ice-covered lake surface, to distinguish the influence of the inhomogeneous landscape on turbulence exchange in the atmosphere-surface system, and exclude the thermal effects of the lake.

2. Field experiment

The measurements were made in the neighborhood of a White Sea Biological Station (WSBS) operated by Moscow State University from 25.01.2017 to 03.02.2017. The measuring system was located on the lake Kislo-Sladkoe, about 2 km east of the station, near the coast of the Kandalaksha Gulf of the White Sea. The lake dimensions are approximately 200 m by 150 m. The southern and western sides of the lake are bordered by vast forest landscapes. The northern and eastern sides are bordered by a forest belt (50–100 m width) separating the lake from the gulf. There are two gaps in the forested areas about 30 m wide each that serve as open surfaces "connecting" the lake with the gulf. The terrain is hilly, while the elevations of topography around the lake do not exceed the average treetop height.

A 6-meter mast was installed in the center of the lake. The fluctuations of temperature and three wind speed components were measured using a WindMaster HS 3D Anemometer (Gill Instruments, UK) at a level of 2 m and a WindMaster 3D Anemometer (Gill Instruments, UK) at levels of 4 m and 6 m with a frequency of 20 Hz. The humidity and air temperature were measured by automatic meteorological stations Davis Vantage Pro 2 (Davis Instruments, USA) at levels of 3 m and 5 m. In addition, a gas analyzer LI-7500 (Li-Cor, USA) was installed at a level of 2 m to calculate moisture fluxes (latent heat) and carbon dioxide. The net radiation was measured using a CMP21 Pyranometer and a CGR3 Pyranometer (Kipp & Zonen, Netherlands). A soil station (Davis Instruments, USA) was mounted at a distance of about 10 m from the mast to measure the temperature of the snow cover on its surface and at the snow-ice interface (10 cm). The temperature at the bottom surface of the ice was assumed to be 0°C, with a measured ice thickness of 32 cm. Measurements on the lake were accompanied by recordings of the temperature profile in the boundary layer of the atmosphere from the surface up to 1 km with a step size of 50 m by a microwave profiler MTP-5 (Attex, Russian Federation).

3. Data processing

The main characteristics of the turbulent interaction of the atmospheric surface boundary layer with the surface are the sensible heat flux (H), latent heat flux (L_sE), and momentum flux (τ). These quantities represent the total fluxes where the effects of molecular exchange are negligibly small in comparison with the vertical turbulent transfer of momentum, heat, and moisture. To determine the turbulent fluxes we used an eddy covariance method, a method of surface energy balance, and a MOST-based gradient method.

3.1. Eddy Covariance method

The direct method, or eddy covariance method, is based on measuring the pulsations of the three wind velocity components u', v', and w', the temperature T', and the specific humidity q' at one point above the surface. The fluxes are calculated as covariances between these variables, i.e. the second mixed moments [5]:

$$\tau = \rho_0\left[\overline{u'w'}^2 + \overline{v'w'}^2\right]^{1/2} = \rho_0 u_*^2, \quad (1)$$

$$H = c_p\rho_0\overline{w'T'}, \quad (2)$$

$$L_sE = \rho_0\overline{w'q'}L_s, \quad (3)$$

where ρ_0 is the reference air density, c_p is the specific heat capacity of air at constant pressure, L_s is the latent heat of vaporization, and the overbar stands for the temporal average. In our case, the fluxes were calculated by 30-minute averaging, which can be assumed to be the optimal averaging time [6]. The EC method involves special corrections to minimize errors, which are mostly of a technical nature. The corrections made to introduce the most significant changes in the heat flux are: time shifting between signals from different sensors, artificial spike removal, "plane approximation" to correct the anemometer tilt [7], Moore spectral correction [8], correction of acoustic temperature and accounting for the effect of air density fluctuations [9]. The EddyUH software package was used to compute the fluxes [10].

The sensible heat and momentum fluxes were calculated at heights of 2, 4, and 6 m with ultrasonic anemometer data at every 30-minute interval. The moisture flux was calculated at a height of 2 m.

3.2. Gradient method

The gradient method is based on the assumption that all statistical characteristics of the temperature, humidity, and wind velocity fields normalized to the corresponding scales of the temperature T_*, humidity q_*, and velocity u_* are described by universal functions of the dimensionless height $\xi=z/L$, where L is the Obukhov length and z is the level of measurements. Specifically, according to MOST the deficits of averaged meteorological variables in the surface layer obey the following relations:

$$\Delta u = u_2 - u_1 = \frac{u_*}{\kappa}\left[ln\frac{z_2}{z_1} - \Psi_u(\xi_2) + \Psi_u(\xi_1)\right], \quad (4)$$

$$\Delta T = T_2 - T_1 = \frac{T_*}{\kappa}\left[ln\frac{z_2}{z_1} - \Psi_T(\xi_2) + \Psi_T(\xi_1)\right], \quad (5)$$

$$\Delta q = q_2 - q_1 = \frac{q_*}{\kappa}\left[ln\frac{z_2}{z_1} - \Psi_q(\xi_2) + \Psi_q(\xi_1)\right]. \quad (6)$$

From (4) - (6) the turbulent fluxes are calculated by:

$$\tau = \rho_0 u_*^2, \quad (7)$$

$$H = -c_p \rho_0 u_* T_*, \quad (8)$$

$$L_s E = -\rho_0 L_s u_* q_*, \quad (9)$$

where κ is the von Karman constant, lower indices “2” and “1” denote the two levels of measurements, $\Psi_i(\xi)$, i=u, T, and q are dimensionless universal functions. Thus, following MOST, one can compute the corresponding fluxes by making measurements at least on two levels.

3.3. Surface energy balance method

The method of surface energy balance is based on the heat conservation law: the heat entering a system equals the sum of the outgoing flux and the change of energy stored in the system. For any surface, the surface heat balance is given by:

$$R_n = H + L_s E + G, \quad (10)$$

where R_n is the net radiation at the surface, and G is the heat flux passing through the surface.

Since shortwave radiation penetrates snow and ice, the heat conduction equation in the snow and ice layers can be written as follows (“s” and “i” indices mean “snow” and “ice”, respectively):

$$\rho_s C_s \frac{\partial T}{\partial t} = \frac{\partial}{\partial z}\lambda_s \frac{\partial T}{\partial z} - \frac{\partial S}{\partial z}, \quad (11)$$

$$\rho_i C_i \frac{\partial T}{\partial t} = \frac{\partial}{\partial z}\lambda_i \frac{\partial T}{\partial z} - \frac{\partial S}{\partial z}, \quad (12)$$

where ρ is the density, C is the heat capacity, λ is the molecular thermal conductivity, T is the temperature, and S is the flux of shortwave radiation. Then we integrate equations (11-12) over the entire thicknesses of the snow and ice, respectively, and sum up the resulting expressions. Assuming that the molecular heat flux and the radiation flux at the ice bottom are negligible in comparison to the surface energy fluxes, we obtain the equation

$$-\lambda_s \frac{\partial T}{\partial z}|_{z=0} = \rho_i C_i \frac{\partial}{\partial t}\int_{h_s}^{h_i} T dz + \rho_s C_s \frac{\partial}{\partial t}\int_{0}^{h_s} T dz - S|_{z=0}. \quad (13)$$

In this experiment, the snow thickness during all measurements was constant, $h_s = 10$ cm, ice $h_i = 32$ cm, the snow temperature was measured at two points, namely, at the surface and at the boundary of snow and ice (0 and 10 cm, respectively), the temperature on the lower ice surface was taken equal to 0°C. The density of ice ($\rho_i = 917$ kg/m^3), the heat capacity of ice ($C_i = 2150$ J/(kg K)) and snow ($C_s = 2150$ J/(kg K)) are the tabular values, and the density of snow ($\rho_s = 100$ kg/m^3) was measured directly. The temperature integral was calculated numerically by the trapezoidal rule. The snow temperature was measured with an interval of 10 seconds.

4. Results

4.1. Comparison of heat fluxes obtained by gradient method and eddy covariance method

Stable stratification in the atmospheric surface layer was observed during the entire experiment, and all turbulent heat fluxes were directed from the atmosphere to the surface. Figure 1 shows the vertical air temperature profile up to 1 km, the sensible heat flux calculated by the gradient method, and the turbulent heat fluxes measured at heights of 2, 4, and 6 m. It can be seen that there are time intervals when the measured fluxes agreed well with the MOST calculations (gradient method), and the other cases when the MOST calculations do not fit the experimental data. Consider both scenarios in more detail:

At some time intervals (for example, those around 26.01 12:00, 31.01 00:00, 01.02 12:00) the results of the gradient method agree well with the direct measurements, albeit being smaller in absolute values. At the same time, the extrema of the heat flux observed during these periods agree well with the increase in the wind speed (up to 5-6 m/s) and with the enhanced difference in temperature between the surface and the atmosphere at the measurement level (not shown). The MOST data demonstrate these heat flux extrema as well, since the flux is proportional to the wind speed and temperature difference between the surface and the atmosphere.

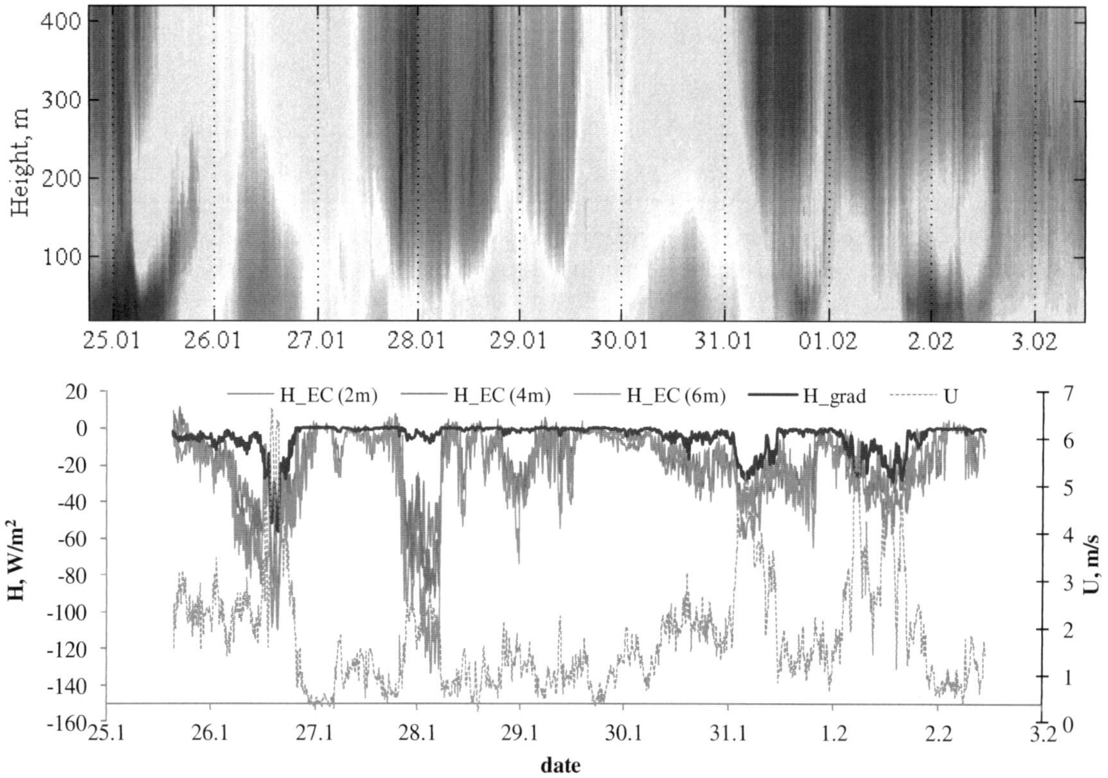

Figure 1. Temperature distribution in the boundary layer of the atmosphere (above) as measured by MTP-5 profiler, fluxes of sensible heat from EC measurements at different levels (H_EC), sensible heat flux by the gradient method (H_grad), and average wind velocity (U).

At other time intervals (for example, those around 28.01 00:00 and 29.01 00:00) there are apparent spikes in the EC-fluxes with no corresponding extrema in the MOST-calculated fluxes. Note that in such cases the wind speed remains small (< 2 m/s), so that the MOST cannot provide a significant increase in the H value. These periods coincide with warm air advection in the middle and upper part of the boundary layer (Figure 1), so that a negative heat flux may form at altitudes about 100 m, following the temperature gradient. We suggest that there may be turbulent diffusion of the heat flux from heights of about 100 m to the surface layer, where the negative heat flux is lower in absolute value. At these times we observe that the heat flux is very sensitive to the wind speed (much more than that calculated by MOST): increasing the speed from 1 m/s to 2 m/s on 28.01 causes the absolute values of fluxes to increase from 40 W/m^2 to 120W/m^2. In this case the flux significantly depends on the height; its modulus increases with height as suggested above. In particular, at 00:00 28.01 the following fluxes were observed: 50 W/m^2 at a level of 2 m, 80 W/m^2 at a level of 4 m, and 120 W/m^2 at a level of 6 m.

A LES simulation of the atmospheric boundary layer over a lake surrounded by forest [3] demonstrated that the absolute magnitude of the heat flux resulting from the temperature differences near the surface decreases with height. In our case, by contrast, it increases with height during the large-flux events. This supports the proposed hypothesis about the role of vertical turbulent transport of heat flux.

One more striking feature seen in Figure 1 is that the measured heat fluxes significantly exceed those computed from MOST, including the periods of a weakly stratified boundary layer (1.02 – 3.02), where the vertical heat flux diffusion is small. A feasible explanation of this is that the effect of landscape inhomogeneity leads to the development of a secondary circulation [11].

4.2. Comparison of fluxes obtained by surface energy balance method and eddy covariance method
Now we consider the total heat flux $H + L_sE$ computed from EC-measurements (z=2 m) in comparison to those calculated with the surface energy balance method. Both series were smoothed to minimize noise (Fig. 2). The span for the moving average is 5.

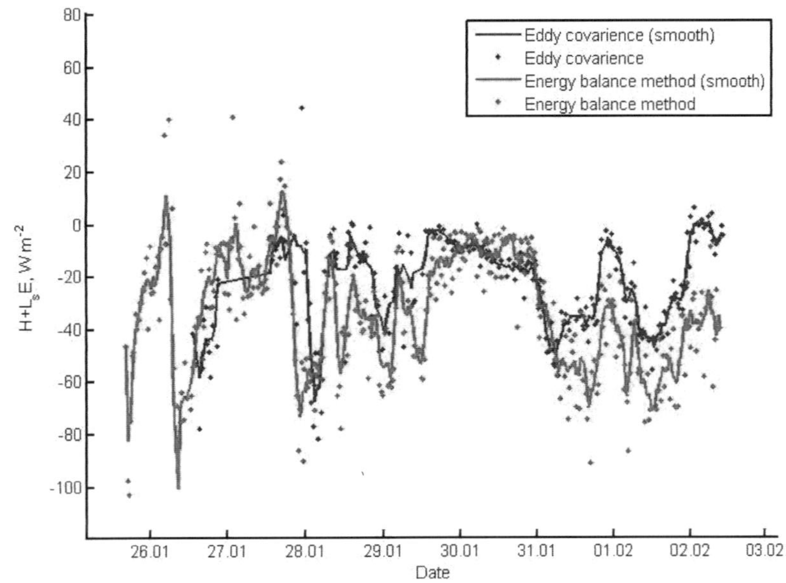

Figure 2. Total heat flux (sensible plus latent) according to direct measurements and calculations from surface energy balance. The span for the moving average is 5.

One can see that the time series correlate well. The greatest difference between the two methods (62.8 W/m^2) is observed on 28.01 during warm air advection above the surface layer. Significant discrepancies are also observed during the last two days of the experiment. In both cases the EC technique underestimates the absolute total heat flux, in line with common experience (as mentioned in the introduction). This is likely a result of the secondary circulations induced by the presence of a forest-lake boundary [11]. The mean difference between the two methods for all period is equal to 16.8 W/m^2. The correlation coefficient of the two time series is equal to 0.6.

4.3. Momentum flux profile
The EC-computed momentum flux time series at each level were divided into two corresponding to different types of heat flux formation, as discussed above:

27.01 20:45 - 28.01 03:15 (MOST fails to predict heat flux).

31.01 02:15 - 31.01 07:45 (MOST predicts heat flux satisfactorily).

The profiles of the averaged momentum flux for both time intervals defined above is shown in Fig. 3.

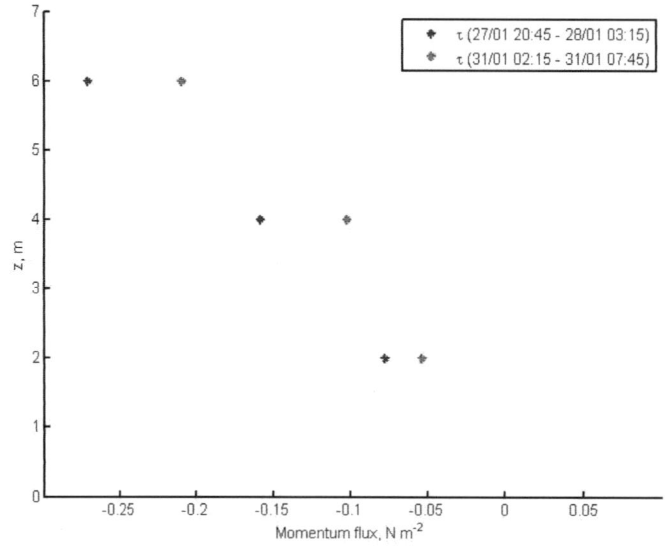

Figure 3. Profile of eddy covariance momentum flux.

The momentum flux increases with height as an approximately linear dependence, and at the surface the momentum flux tends to 0. This agrees well with the LES of inhomogeneous boundary layers [3]. Thus, landscape inhomogeneity is very likely to significantly impact the momentum budget in the surface layer above Kislo-Sladkoe Lake.

5. Conclusions

The turbulence measurements over a frozen lake surrounded by forest demonstrate that eddy covariance fluxes correlate well with those obtained by the surface energy balance method. The tendency of the EC technique to underestimate fluxes derived from the surface energy balance is in agreement with the energy balance closure problem described in [11]. This may be attributed to the secondary circulations developing near the forest-lake boundary. The presence of wind-shear effects at the height of trees demonstrated recently in LES [3] was indirectly supported in our measurements by an increase of momentum flux with height in the surface layer, extrapolating to zero at the surface. However, contrary to LES [3], the sensible heat flux decreased with height most of the time. We suggest that this may be at least partially attributable to the high magnitude negative heat fluxes above the surface layer formed when warm advection occurs at altitudes of ~100 m. During the warm advection events MOST fails to reproduce the sharp increase of negative heat flux in the surface layer. This can be explained by the presence of a vertical transport of heat flux from above not taken into account in the theory. Beyond the warm advection events, the MOST data correspond well to the EC fluxes, however, with some systematic underestimation of the flux magnitude. This is likely to be a consequence of omitting the surface inhomogeneity effects in the MOST calculations.

6. Acknowledgments

This work was supported by RSF grant (17-17-01210).

References

[1] Rouse W R, Oswald C J, Binyamin J, Spence C, Schertzer W M, Blanken P D, Bussieres N and Duguay C R 2005 *J. Hydromet.* **6** № 3 291–305
[2] Downing J A, Praire Y T, Cole J J et al. 2006 *Limnol. Oceanogr.* **51** № 5 2388–97
[3] Glazunov A V and Stepanenko V M 2015 *Izvestia. Atmospheric and Oceanic Physics.* **51** issue 4 351-61
[4] Foken T and Oncley S P 1995 *Bulletin of the American Meteorological Society* **76** 1191-3
[5] Andreas E L, Jordan R E and Makshtas A P 2005 *Boundary Layer Meteorology* **114** 439–60.
[6] Volkov Yu A and Repina I A 2002 *Surface and internal waves in the Arctic seas* ed I V Lavrenova and E G Morozov (Saint Petersburg: Gidrometeoizdat) chapter 11 pp 189-206
[7] Wilczak J M, Oncley S P and Stage S A 2001 *Boundary Layer Meteorology* **99** 127-50
[8] Moore C J 1986 *Boundary Layer Meteorology* **37** 17-35
[9] Webb E K, Pearman G I and Leuning R 1980 *Quarterly Journal of the Royal Meteorological Society* **106** 85-100
[10] https://www.atm.helsinki.fi/Eddy_Covariance/EddyUHsoftware.php
[11] Foken T 2008 *Ecological Applications* **18** № 6 1351-67

Mesoscale atmospheric modelling technology as a tool for creating a long-term meteorological dataset

Vladimir Platonov, Alexander Kislov, Gdaly Rivin, Mikhail Varentsov, Inna Rozinkina, Mikhail Nikitin and Mikhail Chumakov

Lomonosov Moscow State University, Department of Geography, Section of Meteorology and Climatology. 119991, Moscow, Leninskie Gory MGU, 1, GSP-1, phone 8(495)939-29-42
Hydrometcentre of Russia, Moscow, 123242, B. Predtechensky per., 11-13, phone 8(495)255-13-09
A.M.Obukhov Institute of Atmospheric Physics, Russian Academy of Sciences. 119017, Moscow, Russian Federation, Pyzhyovskiy pereulok, 3, phone 8 (495) 951-85-49.
Gazprom VNIIGAZ LLC, Moscow region, 115583, box 130, Razvilka village, 15-1.

vplatonov86@gmail.com, avkislov@mail.ru

Abstract. A detailed hydrodynamic simulation of major meteorological parameters for the last 30 years (1985 – 2014) has been performed for the Sea of Okhotsk and the Sakhalin Island. The regional non-hydrostatic atmospheric model COSMO-CLM was used for this long-term simulation with horizontal resolutions of ~13.2, ~6.6 and ~2.2 km. This dataset was created to help in the investigation of statistical characteristics and physical mechanisms of formation of extreme weather events (primarily wind speed extremes) on small spatio-temporal scales. The detailed meteorological information thus obtained could be used to take into account the coast configuration, mountain systems, and other important mesoscale features of the terrain. This paper describes a proposed downscaling technology for long-term simulations with three "nested domains". The results of verification of the dataset and estimation of extreme wind velocities are presented.

1. Introduction

The hydrometeorological information about the Arctic and Far East regions is presented, apart from poor station data, by reanalysis data (NCEP/NCAR [1], NCEP-CFSR [2], ERA-Interim [3], MERRA [4], NARR [5], ASR [6,7], etc.) and global climate modeling data with a spatial resolution of about tens of kilometers. This resolution may be insufficient for many goals and applications. For example, extreme weather events, especially genesis of extreme wind speeds, are evidently associated with local-scale and mesoscale processes including non-hydrostatic effects. In Kislov et al. [8] it was shown that a global model could not reproduce a significant part of wind extremes distribution, so-called 'dragons' [9]. It is good reason to use more detailed resolution and more precise models and

Content from this work may be used under the terms of the Creative Commons Attribution 3.0 licence. Any further distribution of this work must maintain attribution to the author(s) and the title of the work, journal citation and DOI.
Published under licence by IOP Publishing Ltd

simulations, namely, non-hydrostatic mesoscale regional atmospheric models. There is an additional advantage for use of mesoscale models, specifically for taking into account complex terrain characteristics, such as coastline, mountain topography, detailed surface features, etc.

However, an investigation of extreme weather events and wind speeds could be considered using the long-term simulations and robust statistical estimates of outputs. The combination of high resolution, detailing of mesoscale processes, a long period for analysis and assessments is a good challenge for new research, explanations of physical mechanisms and possible trends of extreme wind speeds. Solution of the corresponding task requires an appropriate model tool, computational resources; many test runs for model configuration optimization, selection of model domains, and development of a general technology, validation and verification of model results.

2. Data and methods

2.1 Model description

The COSMO-CLM model (ver. 5.0) is used as the main tool for the creation of this long-term meteorological archive. COSMO-CLM is the climate version of the well-known mesoscale COSMO model, including some modifications and extensions adapted to the long-term numerical experiments. It is developed by German Weather Service (DWD) and CLM-Community (see CLM Community site http://www.clm-community.eu/, [10, 11]). The COSMO-CLM model is based on the primitive Navier-Stokes equations describing the dynamics of compressible fluid in the moist atmosphere. The model equations are solved on the rotational grid 'latitude-longitude' (λ, φ) with a pole tilt. It helps to minimize the problem of meridians convergence over the pole. The numerical scheme is realized on Arakawa C-grid, and the vertical coordinate is a hybrid Gal-Chen coordinate.

The standard configuration of the COSMO-CLM model was applied with the Runge-Kutta integration scheme with 5th advection order. The height-based hybrid Gal-Chen coordinate [12] is given as an analogue of the σ-coordinate from the surface up to Z_0 level and as the Z-coordinate above the Z_0 level. The Ritter and Geleyn radiation scheme [13] is based on the δ two-stream version of the radiation transfer equation. The precipitation formation described by a bulk microphysics parameterization, Tiedtke mass-flux schemes with equilibrium closure based on moisture convergence are used for moist and shallow convection [14]. Turbulence is described by a prognostic TKE-based scheme, with 2.5 order closure [15]; the spectral nudging technique [16] and Smagorinsky diffusion included. A full description of the COSMO Model physics, dynamics and parameterizations is available on (http://www.cosmo-model.org/content/model/documentation/core/default.htm).

2.2 Experiments and downscaling technology description

The COSMO-CLM model runs were executed using ERA-Interim reanalysis [3] data (~0.75^0 resolution) as driving conditions over the base domain (~13.2 km resolution) for the 1985 – 2014 period (i.e. 30 years). Additionally, many external parameters data (e.g., land/sea mask, roughness length, leaf area index, etc.) came from EXTPAR v.3.0 tool provided by CLM-Community (www.clm-community.eu/index.php?menuid=174&reporeid=260) over the base domain (data from EXTPAR are provided in Table 1). All these data went into preprocessing routine int2lm and were interpolated onto the domain grid.

Generally, the 'nested domains' technique was realized through three nested computational areas, with 13.2, 6.6, and 2.2 km resolutions. As a result, more than 20 detailed meteorological fields (see Appendix) within the 30-year period with 1-hour temporal resolution over the Okhotsk Sea and Sakhalin region were obtained. Initially, the model run was executed over the starting domain ('base domain') with a raw 13.2-km resolution through the whole period. After that, these outputs were used as initial and boundary conditions for interpolation and model run over the next domain, with 6.6 km resolution. And, finally, in the same way, we have executed model runs over the 2.2 km resolution domain with the driving conditions from the previous domain's outputs. However, some shortcomings of the downscaling method should be noted. The first one is associated with occurring artifacts and spurious effects at the domain boundaries. The second one is the bias associated with coarse resolution, which the regional climate model also adds by its own errors to the output data. Nevertheless, using the 'nested domains' scheme with finer resolution and reasonable choice of boundaries could allow us to reduce these errors and biases.

Table 1. External parameters list provided by EXTPAR tool and required for the COSMO-CLM model run initialization.

Parameter	Name	Units
HSURF	Geometrical height	m
Z0	Roughness length	m
LAI_V	Leaf area index, vegetation period	
LAI_R	Leaf area index, non-vegetation period	
FR_LAND	Land cover fraction	1
PLCOV_V	ground fraction covered by plants, vegetation period	1
PLCOV_R	ground fraction covered by plants, non-vegetation period	1
ROOTDP	Root depth	m
SOILTYP	Soil type	-
T_CL	Soil temperature	K
FR_LAKE	Lakes fraction cover	1

The 'spectral nudging' technique was additionally applied over the base domain in order to control the model behavior and link to the real atmospheric dynamics. This technique assimilates the large-scale components of atmospheric circulation from reanalysis data not only on the boundaries, but also within the whole model domain. It limits the possible model retreat from the real conditions. It is based on the two-dimensional Fourier decomposition of reanalysis and regional model fields and the succeeding adjustment of simulation results. The temperature, wind speed, geopotential, and pressure were assimilated using this spectral nudging technique. Taking into account that the global control of atmospheric circulation is executed by the large-scale systems in the middle and upper troposphere, we have assumed 500 km and more as the spatial scale, and 850-hPa pressure level as the lowest for assimilation.

Practically, this computational scheme was realized with model runs for a period of several months (from 3 – 4 up to a year). Such duration was used because of restrictions of computational resources and data storage volumes, as well as a risk of technical crashes of experiments during the continuous runs for longer periods. All simulations were executed on 288 cores of the RSC "Tornado" supercomputer system in the Main Computer Center of Roshydromet with a peak performance of 35 Tflops. 30-year runs over 13.2 and 6.6 km domains together used ~2,400 hours of the CPU time. Additionally, many hours consumed test runs, debugging experiments, and simulations over the 2.2-km domain (for the most extreme wind speed events), approximately 200 – 300 CPU hours.

2.3. Selection of domains' boundaries

The selection of domains for simulations is one of the most important stages of preparation to simulations. There are many competing considerations to take into account. First, it is necessary to find a compromise between the limitations of computational resources and the size of the territory. Second, it is not recommended to draw the domain boundary near large mountain ranges, because it affects gravitational boundary waves propagation. Third, the boundaries of the next nested domain should be shifted inside from the previous ones by 4 – 10 model grids, i.e. by 20 – 50 km. Ultimately, trying to take into account all these considerations, we have selected the following boundaries, as shown in Figure 1. The main characteristics of these domains are listed in Table 2.

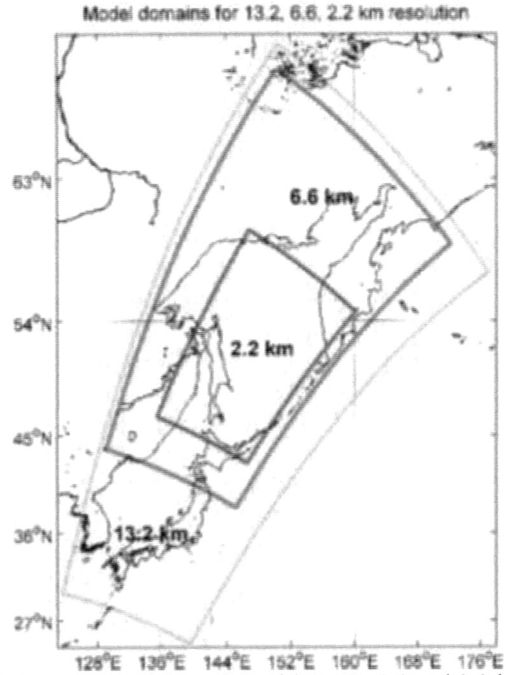

Figure 1. Map of boundaries of model domains with 13.2, 6.6 and 2.2 km horizontal resolutions.

Table 2. Main characteristics of model domains.

Characteristics	13.2 km domain	6.6 km domain	2.2 km domain
Longitude of tilted pole	110^0 E	110^0 E	110^0 E
Latitude of tilted pole	60^0 N	60^0 N	60^0 N
Total number of grid points	145*355 = 51475	228*525 = 119700	300*500 = 150000
Horizontal resolution, grad (km)	0.12^0 (~13.2 km)	0.06^0 (~6.6 km)	0.02^0 (~2.2 km)
Time step, sec	120	60	20
Number of model levels in atmosphere	40	40	50
Number of model levels in soil	9	9	9
Source of initial and boundary conditions	ERA-Interim (~0.75^0)	COSMO-CLM 13.2 km	COSMO-CLM 6.6 km

An important reason to select boundaries for the 'base' domain (13.2 km resolution) was accounting for the synoptic-scale features of atmospheric circulation over the given region, including the monsoon system and the associated cyclonic activity, both in summer typhoons and winter polar fronts. Therefore, the eastern boundaries were extended farther to the Pacific Ocean. It allowed to catch out severe cyclones at early stages of its formation and reproduce its evolution better. A realistic simulation over the 'base' domain was important to get adequate driving conditions for the 6.6 km domain runs. The boundaries of the 6.6 km domain are only slightly smaller than the 'base' one, contouring the Okhotsk Sea. This boundary seems to be sufficient for a good reproduction of the wind speed climatology over the region. The 2.2 km domain covers the Sakhalin Island, parts of the Okhotsk Sea, the Kamchatka peninsula, and the Japan Sea. Since the main goal of this domain was to simulate the extreme wind speeds near the Sakhalin Island, this domain was enlarged over the Okhotsk Sea in order to resolve and reproduce many mesoscale cyclonic features contributing to the formation of extreme winds.

3. Results and discussion

3.1. Verification of dataset

The verification of the obtained dataset was performed with observation data obtained from the "Hydrometcenter database" and www.rp5.ru archive on temperature, wind speed, and wind gusts (U_g > 10 m/s). The stations located in less than 100 km from the coast and 500 m of absolute height were sorted out. Estimates of the model quality were obtained over three verification areas (see Figure 2): 1) Okhotsk sea, adjacent water and coastal areas (124 stations), 2) Sakhalin Island, strait of Tartary, the southern part of Okhotsk Sea (50 stations), and 3) the northern part of the eastern Sakhalin coast (6 stations). The primary comparison was made for 2014 because of the most thorough observational data archive. We have considered four periods for assessment – January-March (JFM), April-June (AMJ), July-September (JAS), and October-December (OND) of 2014.

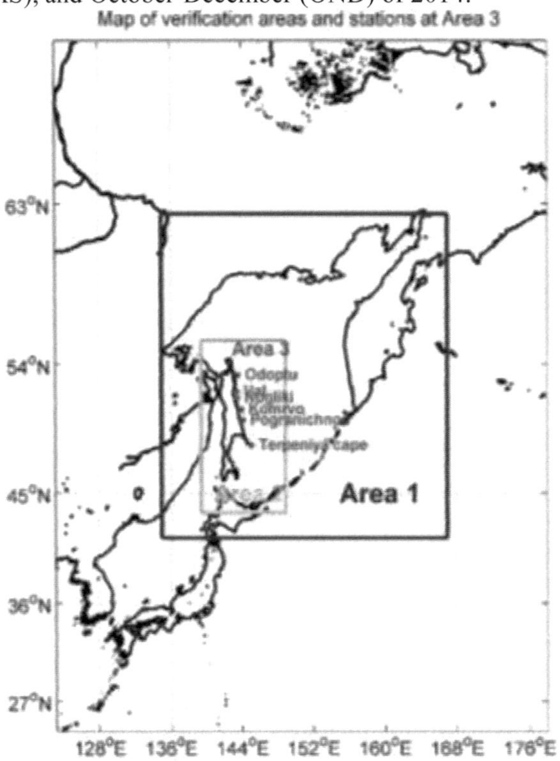

Figure 2. Map of verification areas 1, 2 and stations in area 3.

We applied a specific technique for verification of the modelled temperature and wind speed. For each station, that model grid for comparison was defined which had the minimal RMSE with observations. The corresponding model grid was found over the 25 x 25 km square around the station, i.e. 3 x 3 model grids with a 13.2-km resolution and 5 x 5 model grids with a 6.6 km resolution. Moreover, stations having more than 5% observation lacunas were rejected.

Summarizing the verification results, the mean temperature errors are about 0.5 ^{0}C, while the RMSE reaches 2 - 3 ^{0}C. The wind speed is overestimated by the model over inlands (RMSE is up to 2 m/s). It's noteworthy that wind gusts were reproduced by the model rather well (the ME was up to 1 m/s, the RMSE was 2 – 3.5 m/s, and the correlation coefficients were 0.8 and more), despite a fairly simple algorithm [17]. Additionally, the most noticeable errors are for Area 1, 2 the errors are less for Area 2, and much less for Area 3.

For several stations, errors may be high, but it could be explained by errors in the model land-mask and its inconsistences with the real surface and terrain in these cases. Seasonal courses of the parameters and error were revealed. RMSE for temperature over Area 1 is maximal in winters and minimal in summers. This is associated with overestimation of temperature by the model during strong freezings and, consequently, stable stratifications near the surface. For the Sakhalin's eastern coast, the situation is vice versa. The correlation coefficients for temperature are more than 0.9 during the entire year; therefore, the model reproduces its synoptic and daily variability correctly. The largest RMSEs for wind speed in the OND period, i.e. the 'storm season', are associated with maxima absolute values in this time. Thus, the lower correlation coefficients during the summer season could be explained by the frequency of convective movements. The transition from the resolution of 13.2 km to 6.6 km does not provide any significant improvement, but one should note a slight decrease of the errors and increase of the correlation coefficients. Also, a decrease of the errors range for parts of the seasons, parameters and areas was observed.

As for the spatial distribution of errors, maximal temperature errors are observed at the coastal stations (mostly in summer) or in inner lands (mostly in winter). The summer errors could be explained by a strong contrast between the sea and land and, hence, by the complexity in reproduction of the local land-sea interactions. The winter errors are linked to the well-known overestimation of turbulent mixing in the boundary layer by the COSMO-CLM model in stable stratification conditions that leads to rising of the model temperature.

The spatial distribution of wind speed errors is simpler. They are larger over the inlands, because of inconsistency between the real and model-defined roughness length in the land-sea mask for several stations. The errors for the coastal stations are smaller due to a less number of factors determining the wind speed.

Verification for the 2.2-km domain was performed only for extreme wind situations listed in Table 3. Generally, comparison characteristics for these short periods of extremes are not worse than analogous ones for seasons. Therefore, the model could reproduce both the background seasonal variability and its dynamics during the extreme winds with the same quality. Using finer resolution slightly improves the model results, especially by transition to 2.2 km.

3.2. Synoptic analysis of extreme wind situations

Extreme wind speed situations were sorted out from the given archive (15 cases listed in Table 3) and analyzed according to the synoptic processes. Two main types of synoptic situations were identified as favorable for the genesis of extreme winds. All selected storms were observed during the cold season. The synoptic features leading to extreme wind speeds were associated with a strong thermal gradient between the cold continental air mass over Eastern Eurasia and the marine polar air masses over the Pacific and Okhotsk Sea during winter. It leads to strengthening of the cyclonic activity at the polar and arctic fronts over the Okhotsk Sea and Sakhalin Island. All selected extreme winds were of northern directions and caused by cyclones.

The first type (most frequent) was associated with developing of cyclones over the Primorsky Kray or Japan Sea (example for 28.01.1989, 07 UTC, Figure 3a). Then it intensified, moved to the west or north coast of the Sakhalin Island, crossed it, and came to the Okhotsk Sea. Here, in the rear of the cyclone, the wind speed usually intensified again. The intensity and duration of extreme winds cases varied from a few hours to a day and 25 to 35 m/s.

Table 3. Dates of extreme synoptic situations over the Okhotsk Sea region and maximal wind speed values (m/s) reproduced by COSMO-CLM model.

№	Date	Wind speed, m/s		
		13.2 km	6.6 km	2.2 km
1	25.03.1987	29.7	31.6	31.1
2	28.01.1989	32.3	34.3	34.3
3	19.12.1989	28.6	30.0	31.8
4	8.11.1995	27.0	26.7	27.0
5	1.02.1996	28.3	29.2	28.9
6	14.02.1996	32.8	34.2	34.0
7	28.02.1999	33.3	35.2	35.5
8	14.11.1999	27.4	25.3	28.7
9	11.01.2000	28.3	29.3	29.6
10	12.01.2001	30.8	30.3	30.7
11	5.12.2002	27.7	30.0	29.0
12	8.01.2005	29.8	28.7	30.6
13	27.02.2006	28.9	29.0	30.6
14	3.02.2014	27.7	25.5	26.7
15	6.03.2014	24.9	24.8	26.8

The second type was characterized by the penetration of intensified cyclones from the Pacific Ocean through the Kamchatka peninsula, Kuril or Japan Islands, especially Hokkaido (example for 19.12.1989, 11 UTC, Figure 3b). Next, the deepening cyclones crossed the Okhotsk Sea and got close to the Sakhalin Island, causing extreme winds offshore. This group was characterized by less deep but larger cyclones.

A comparison of the reproduction of these extreme situations at different model resolutions showed that on 2.2 km the wind speed maximums are slightly higher (the added value is 2 – 3 m/s) and its spatial distribution is more spotted compared to rawer resolutions. In most cases using a finer resolution leads to decreasing of the mean speed and gusts errors. At the same time, these features are not so evident by transition from 13.2 to 6.6 km, in contrast to the transition from 6.6 km to 2.2 km. It follows that 6.6 km resolution does not provide a significant improvement of the model results due to a bad resolving of the 'grey zone' processes, obviously. However, the 2.2 km resolution improves the reproduction of extreme winds in the surface layer because of explicit resolving of deep convection, detailed terrain, coastline, and turbulent motions description.

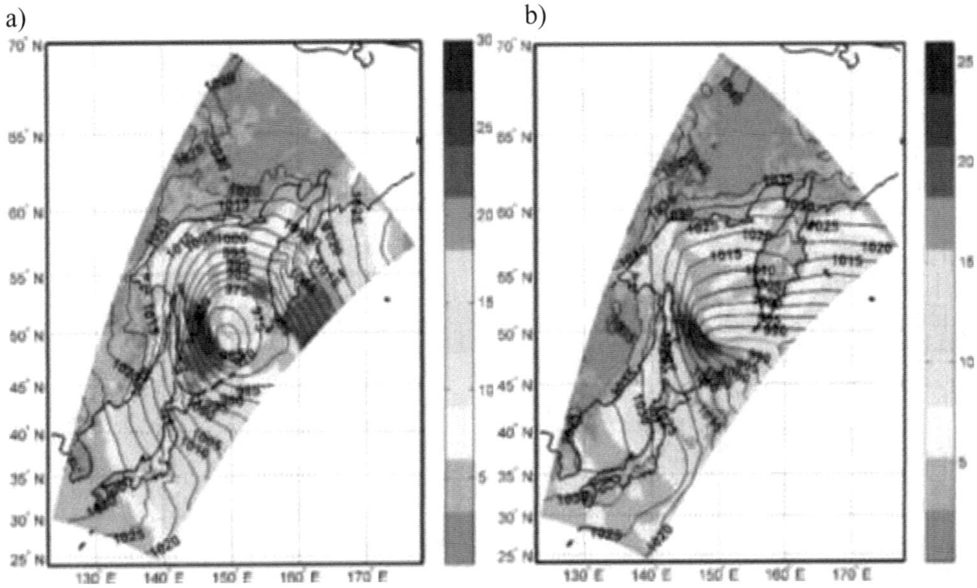

Figure 3. Sea level pressure (hPa) and wind speed (m/s) 28.01.1989, 07 UTC (a) and 19.12.1989, 11 UTC (b) reproduced by COSMO-CLM model on the 13.2 km domain.

4. Conclusions and recommendations for future research

It is for the first time in the Russian Federation that a technology of detailed regional modelling of atmospheric processes was realized on long-term timescales. Overall, the analysis showed that the used downscaling technique with applying the COSMO-CLM model reproduced the meteorological conditions, spatial distribution, seasonal and synoptic variability of temperature and extreme wind speed for the study area with approximately the same adequate quality. Some relations between the reproduction quality of mesoscale atmospheric circulation features and the horizontal resolution of the model were revealed. In particular, it was shown that ~6.6 km resolution does not provide any significant improvement in comparison to ~13.2 km resolution, whereas ~2.2 km resolution provides an appreciable improvement of the quality.

The obtained high-resolution dataset will be used for a full and comprehensive analysis of the physical mechanisms of extreme weather events, the reproduction quality of hydrometeorological fields, their statistical estimates, climatological trends, spatial distribution of a large variety of hydrometeorological parameters on diurnal, seasonal, and interannual timescales, using this information for a detailed environment state assessment and many other objectives.

Acknowledgements

The authors wish to thank the director of "Hydrometcenter of Russia" Roman M. Vilfand and the director of "Main Computational Center of Roshydromet" Sergey V. Lubov for their assistance and help in this work.

This research was supported by Russian Science Foundation (grant no. 14-37-00038), Russian Foundation for Basic Researches (grant RFBR no. 16-05-00509) and state project no. AAAA-A16-116032810086-4.

Appendix

List of main output variables:

- Sea level pressure (hPa)
- Air temperature 2 and 10 m (^{0}C)

- Potential air temperature 2 and 10 m (^{0}C)
- Surface temperature (^{0}C)
- Specific humidity 2 and 10 m (g/kg)
- 10 m zonal and meridional wind (m/s)
- 10 m wind speed maxima during 1 hour (m/s)
- 10 m wind gust during 1 hour (m/s)
- Hourly snow and rain precipitation (kg/m^2, both)
- Radiation (direct and diffuse) shorter and longer than 700 nm (W/m^2, both)
- Downward shortwave radiation (W/m^2)
- Upward shortwave radiation (W/m^2)
- Downward and upward longwave radiation (W/m^2)
- Sensible and latent heat flux (W/m^2)

References

[1] Kalnay E *et al.* 1996 The NCEP/NCAR 40-year reanalysis project *BAMS* **77** 3 437 – 71
[2] Saha S *et al.* 2010 The NCEP climate forecast system reanalysis *BAMS* **91** 1015
[3] Dee D P *et al.* 2011The ERA-Interim reanalysis: Configuration and performance of the data assimilation system *Quart. J. RMS* **137** 656 553 – 97
[4] Rienecker M M *et al.* 2011MERRA: NASA's modern-era retrospective analysis for research and applications *J. of Climate* **24** 14 3624 – 48
[5] Mesinger F *et al.* 2006North American Regional Reanalysis *BAMS* **3** 343 – 60
[6] Bromwich D H *et al.* 2009 Data assimilation challenges for high-resolution reanalysis in the Polar regions: The Arctic System Reanlaysis *10th Annual WRF Users' Workshop* 1 – 4
[7] Bromwich D H, Wilson A B, Bai L-S, Moore G W K, Bauer P A 2016 Comparison of the regional Arctic System Reanalysis and the global ERA-Interim Reanalysis for the Arctic *Quart. J. of RMS* **142** 644 – 58
[8] Kislov A, Matveeva T, Platonov V 2015 Wind speed extremes in Arctic area 2015 *Fund. and Appl. Clim.* **2** 63–80 (In Russian)
[9] Kislov A, Matveeva T 2016 An extreme value analysis of wind speed over the European and Siberian parts of Arctic region *Atm. and Clim. Sci.* **6** 205 – 23
[10] Böhm U, Kuecken M, Ahrens W, Block A, Hauffe D, Keuler K, Rockel B, Will A 2006 CLM – the climate version of LM: brief description and long-term applications *COSMO Newsletters* **6** 225–35
[11] Rockel B, Will A, Hense A 2008 The regional climate model COSMO-CLM (CCLM) *Met. Zeit.* **17** **4** 347–48
[12] Gal-Chen T, Somerville R C J 1975 On the use of a coordinate transformation for the solution of the Navier-Stokes equations *J. of Comput. Physics* **17** 2 209 – 28
[13] Ritter B, Geleyn J F 1992 A comprehensive radiation scheme for numerical weather prediction models with potential applications in climate simulations *Mon. W. Rev.* **120** 2 303 – 25
[14] Tiedtke M 1989 A comprehensive mass flux scheme for cumulus parameterization in large-scale models *Mon. Wea. Rev.* **117** 8 1779 – 800
[15] Herzog H-J, Vogel G and Schubert U 2002 LLM – a nonhydrostatic model applied to high-resolving simulations of turbulent fluxes over heterogeneous terrain *Theor. Appl. Clim.* **73** 67 – 86
[16] Von Storch H, Langenberg H, Feser F 2000 A spectral nudging technique for dynamical downscaling purposes *Mon. Wea. Rev.* **128** 10 3664 – 73
[17] Schulz J-P, Heise E 2003 A new scheme for diagnosing near-surface convective gusts *COSMO Newsletter* **3** 221 – 5

IOP Conf. Series: Earth and Environmental Science **96** (2017) 012005 doi:10.1088/1755-1315/96/1/012005

A new earth's climate system model of intermediate complexity, PlaSim-ICMMG-1.0: description and performance

G Platov[1,3], V Krupchatnikov[2,3,4], Yu Martynova[2,4], I Borovko[1] and E Golubeva[1,3]

1 Institute Computational Mathematics and mathematical geophysics SB RAS
2 Institute of Monitoring of Climatic and Ecological Systems SB RAS, Tomsk, 634055, Russia
3 National Research Novosibirsk State University, 630090, Novosibirsk, Russia
4 Siberian research hydrometeorology institute of Roshydromet, 630099, Novosibirsk, Russia

e-mail: vkrupchatnikov@yandex.ru

Abstract. We describe a new intermediate complexity model (PlaSim-ESM-ICMMG-v.1.0) constructed by coupling the Planet Simulator [1-2] with ocean and sea-ice models. We demonstrate the results of climate simulation using PlaSim-ICMMG-v.1.0 considering global fields of surface air temperature, precipitation, sea surface temperature, and ocean circulation, and make comparisons with the results obtained using the original PlaSim version. PlaSim-ICMMG-v.1.0 reproduces the main features of the climate system reasonably well and demonstrates that it is useful for climate system modeling. Due to the rapid warming in the Arctic, there are major challenges associated with the mechanisms that regulate the dynamics of weather in the mid-latitudes. The new earth's system model of intermediate complexity PlaSim-ICMMG-v.1.0 can be used to deal with these challenges.

1.Introduction

Intermediate-complexity models are models describing the dynamics of the atmosphere and/or ocean in less detail than conventional General Circulation Models (GCMs). At the same time, they go beyond the approach taken by atmospheric Energy Balance Models (EBMs) or ocean box models by using sophisticated parameterizations of the unresolved flow or by explicitly resolving the equations of geophysical fluid dynamics, although at a coarser spatial resolution. Being computationally fast, intermediate-complexity models are capable of treating slow climate variations. Hence, they often include components of the climate system that are associated with long-term feedbacks like ice sheets, vegetation, and biogeochemical cycles. There exist many various approaches to building such a reduced model, resulting in a 'spectrum of Earth's system Models of Intermediate Complexity closing the gap between EBMs and complex GCMs [1-10].

Over the last few decades, the Arctic region has been warming much more than in the global average, primarily in winter, while the Arctic sea-ice extent has been decreasing dramatically [11-14]. Many mechanisms are at work in the 'Arctic amplification'. The existing positive snow and sea-ice albedo feedback plays a significant role in amplifying the warming signal [15]. The albedo feedback operates in summer when solar radiation is maximal. When the sea ice is shrinking and the water surface is opening, the warming due to the absorbed shortwave radiation can be large and enhances the sea-ice loss through lateral melting. Beyond the atmospheric heat transports, the high-latitude response to the greenhouse forcing involves an anomalous ocean heat transport into the Arctic. This occurs even if a weakened meridional overturning circulation (MOC) diminishes heat transport in the lower

Content from this work may be used under the terms of the Creative Commons Attribution 3.0 licence. Any further distribution of this work must maintain attribution to the author(s) and the title of the work, journal citation and DOI.
Published under licence by IOP Publishing Ltd

latitudes. In addition, the ocean can act as a reservoir for the heat gained in summer when the sea ice retreats, storing it in winter months when the sea ice comes back.

The sea ice reduction is accelerating in the Barents and Kara Seas. The downward longwave radiation is an essential element of the sea ice reduction, but can only be sustained by an excessive upward heat flux from the sea surface exposed to air in the region of sea ice loss. The increased turbulent heat flux is used to increase the air temperature and specific humidity in the lower troposphere, which, in turn, increases the downward longwave radiation. This feedback process is clearly observed in the Barents and Kara Seas in the reanalysis data. A quantitative assessment reveals that this feedback process is amplifying at a rate of ~8.9 % every year in 1979-2016. Based on this estimate, the sea ice will completely disappear in the Barents and Kara Seas by around 2025. The excessive heat flux is necessary for the maintenance of this feedback process; a similar mechanism of sea ice loss is expected to take place in the sea-ice covered polar region when the sea ice is not fully recovered in winter.

In addition to these processes, the warmed ocean mixed layer delays the sea-ice growth, and, thus, affects the wintertime surface temperatures through a thinner ice pack. Since the Arctic atmosphere is stably stratified by thermal inversion at the surface, any warming that occurs there does not go far up into the troposphere.

There is a challenge whether the Arctic amplification affects the weather events in the midlatitudes [15-22]. What are the mechanisms? The rapid Arctic warming is expected to affect the weather patterns in the northern hemisphere, but how, where, and when is the topic of a study. Extreme weather events have been observed more frequently in the recent decades. A new earth's system model of intermediate complexity (PlaSim-ICMMG-v.1.0) has been constructed to study these problems as well.

The Integrated Earth's system model PlaSim-ESM-ICMMG-v.1.0 has been developed as a modular framework allowing a spectrum of intermediate-complexity Earth's system models to be created by selecting different options for the various climate and carbon cycle components. The plaSim-ESM-INMMG-1.0 model is capable of integration over multi timescales long runs. The framework has been designed to be modular to facilitate the coupling of more complex component modules as the available computing power increases.

2. Description of Earth's Climate System Model Plasim-ECS-ICMMG-1.0

2.1. PlaSim Version

PLASIM is a climate system model of intermediate complexity [1].

The PlaSim atmospheric dynamics is solved using a spectral transform method formulated for temperature, specific humidity, log surface pressure, divergence, and vorticity, and is described in detail in [2]. The shortwave radiation scheme used separates solar radiation into two bands, $\lambda < 0.75$ μm (with cloud scattering, ozone absorption, and Rayleigh scattering) and $\lambda > 0.75$ μm (with cloud scattering, cloud absorption, and water vapor absorption). The longwave radiation scheme uses a broadband emissivity method, with the (greenhouse gas) effects of water vapor, carbon dioxide, and ozone included in the calculation of emissivity. The ozone concentration is prescribed with an analytic spatio-temporal distribution. The cloud emissivity is calculated from the cloud liquid water content. The fractional cloud cover is diagnosed from the relative humidity (stratiform clouds) and from the convective precipitation rate (convective clouds). The other parameterized processes are largescale precipitation, moist convection (both cumulus and shallow), dry convection, boundary layer heat fluxes, vertical diffusion, and horizontal diffusion. A simple terrestrial dynamic global vegetation model land surface scheme, SimBA [2] is used to obtain the land surface variables.

The local runoff is transported to the ocean by a river transport scheme with linear advection.

The ocean and sea ice modules provide sea surface temperatures T_{sea}, sea ice distributions c_{ice}, and surface temperatures over the sea ice T_i. The ocean mixed layer heat flux (Qo) represents the oceanic transport and the deep water. The slab ocean model consists of a prognostic equation at each ocean point for the oceanic mixed-layer temperature T_{mix}. [2] (Planet Simulator -Reference Manual Version 15 by F. Lunkeit et al., 2007).

2.2. PlaSim-ESM-ICMMG-1 Version

2.2.1. Ocean and Ice Models. As a basis for describing the oceanic block of the climate system, a model of the Arctic and North Atlantic developed at the Institute of Computational Mathematics and Mathematical Geophysics SB RAS is taken. This model was developed as a result of a significant improvement [24] in an earlier model of the World Ocean [25, 26]. The model is based on the traditional equations of ocean thermodynamics in curvilinear orthogonal coordinates using the hydrostatic and Boussinesq approximations. To specify surface boundary conditions, the "rigid lid" approximation is used. After vertical integration of the equations of fluid motion, the solution reduces to finding an integral component (the barotropic mode) and a deviation from it (the baroclinic mode). For the barotropic mode, the equations of motion are formulated in terms of an integral stream function. When integrating over time, a hybrid explicit-implicit scheme is used, splitting into physical processes and spatial coordinates. A more detailed numerical solution is described in [27]. In this version of the numerical model, the advective-diffusion operator is split into two operators describing the advection and diffusion process separately. To approximate the horizontal and vertical advection, a numerical scheme, ULTIMATE QUICKEST, proposed in [28] is used. The parametrization of vertical turbulence is based on the Richardson criterion and the vertical mixing procedure with the formation of a layer of a vertically homogeneous distribution of hydrodynamic characteristics. More details of the procedure can be found in [26]. Restoring of the surface salinity is not used.

The integration of the ocean module is carried out in conjunction with a model of thermodynamics of elastic viscous-plastic ice, CICE 3.1 [29], which is a modification of an earlier viscous-plastic Hibler's model [30].

As external characteristics determining the conditions for the development of sea ice, the following is used: the wind velocity, ocean surface velocity, specific humidity, density and temperature of the surface air, incoming short-wave and long-wave radiation, precipitation rate in the form of rain and snow, temperature and salinity of the ocean surface, and ocean surface tilt. The ice model calculates the following characteristics at the ice-water interface: the penetrating short-wave radiation, heat flux, salt and fresh water, and the frictional stress between water and ice.

In an earlier stand-alone version of the ice-ocean model, the characteristics of the lower atmosphere necessary for calculating the heat fluxes, moisture, wind friction stresses, and the downstream radiation fluxes were taken from the available data, for example, from the NCEP/NCAR reanalysis data. Within the climatic model system, the supplier of these data will be an atmospheric module. The initial distribution of the temperature and salinity fields corresponds to the climatic data of the Polar science center Hydrographic Climatology (PHC) [31] for winter. The initial ice characteristics were obtained as a result of a preliminary 10-year calculation with NCEP/NCAR 1948-1957 forcing.

2.2.2. Model Grid. The numerical grid of the model from 90S to 65N coincides with the lines of parallels and meridians with a resolution of $1\times1^{\circ}$. At a latitude of 65N, this grid goes into a grid for the polar region constructed as a result of rotating a spherical coordinate system and re-projecting its hemisphere to a region above 65N (see [32] for more details). The horizontal resolution of the polar grid varies from 14 to 55 km with an average grid step of about 20 km. In the vertical direction, the grid consists of 38 levels with a maximum resolution of 5 m in the upper 20-meter layer. The minimum depth on the shelf is 20 m. The position of the shoreline is as close as possible to the geographic position taking into account the horizontal resolution. At the grid nodes where the depth was less than 20 m, the depth was artificially assumed to be 20 m.

The model domain includes the entire World Ocean. Some of the straits were artificially expanded in order to correspond to the model resolution.

2.2.3. River runoff. The river flow is calculated independently by a land model. Numerically, the river runoff is accounted for by additional precipitation, which is set at the grid points close to the geographic location of the river mouths.

2.2.4. Solid boundaries. The boundary conditions at the bottom and at the solid lateral boundaries ensure the absence of heat flows and salt flows through these boundaries. The bottom friction is proportional to the square of the near-bottom flow velocity; a no-slip condition is set on the lateral boundaries.

2.2.5. Parallelization. The modern problem of climate and climate systems research is increasingly associated with numerical modeling using high spatial resolution of the physical processes. The possibilities of solving it within the framework of serial numerical modeling have largely been exhausted. The creation and development of parallel models is an extremely topical activity. However, splitting into separate parallel modules is not enough, and parallelization of each module is required. The model of ice thermodynamics and ice drift CICE is based on explicit numerical schemes and is already parallelized based on a domain decomposition. The ocean model described above originally used implicit numerical schemes, which causes some difficulties in parallelization. Therefore, it was decided to translate the algorithms to use only explicit schemes horizontally and implicit ones vertically. This approach was implemented and has demonstrated its effectiveness in a number of test experiments.

3. Interaction between the climate system components

The model of the climatic system of the Arctic Ocean developed at ICMMG is a set of parallelly working and interacting modules responsible for the simulation of various components of the climate system. Until recently these were: an ocean circulation module, an ice thermodynamics module, and a drift module. The atmospheric module was organized in the form of a data model, i.e. in the form of a

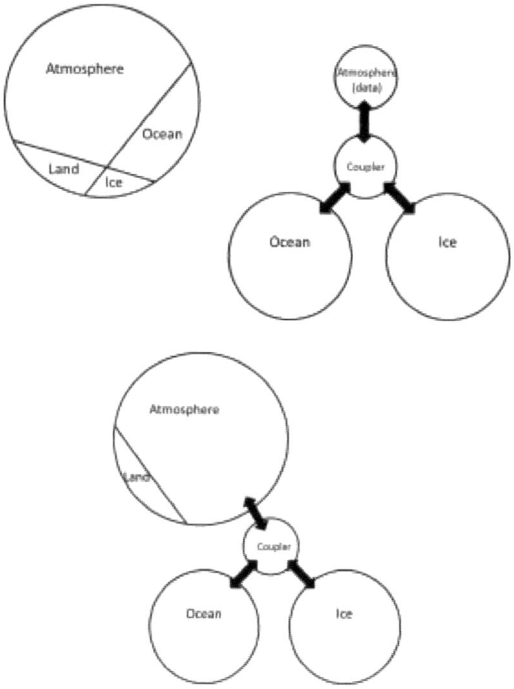

(a)

(b)

Figure 1. Graphical representation of the climate system components in the original version (a) and in the final variant (b).

set of programs for reading reanalysis data and their subsequent interpolation to the model grid. In addition, the software package includes a separate module called "coupler" which provides synchronization and interaction of the components (Figure 1a).

On the other hand, the PlaSim model already contains ocean and ice modules, whose quality does not fully correspond to the complexity of the tasks being solved and requires the replacement by modules of greater physical validity.

However, it is not just a question of replacing one set of subprograms by another set. In PlaSim, a grid that is common for all components is used, on the basis of which parallelism is performed by the method of domain decomposition. The use of a common grid is not advisable. Therefore, a variant was implemented where, instead of accessing the existing ocean and ice modules, a coupler module was called with necessary data exchange to form the flux fields between the system components (Figure 1b).

3.1. Coupler

In the original version, a coupler module was implemented in such a way that all components were supposed to work on the same grid. However, this approach, although it allows saving the calculation time due to the lack of interpolations, is not convenient. The grid used in the atmospheric module must be in a spherical coordinate system because of the need for spectral decomposition of the solution. The use of such a grid in the ocean model is inexpedient, because no spectral expansion is used here, and a direct numerical solution will be difficult due to a singularity of the coordinate system in the north pole region. Therefore, the ocean model grid should remain tripolar (such as described above). Moreover, this grid should allow an increase in the horizontal resolution, in contrast to the grid for the ice model, where the assumption of a statistical distribution of ice (used in CICE 3.1) is valid only with a rough resolution. In addition, the World Ocean contains areas where ice is never formed, and it would be optimal to exclude them from the sea-ice model domain. Therefore, a new revision of the coupler was formed based on the assumption that each of the interacting modules has its own grid and solution area, and it is required to provide not only their time synchronization, but also rapid interpolation from one grid to another. The numerical integration of the entire system in time should be organized so that the slowest modules work smoothly, i.e. the waiting time for a response from the coupler should be minimized.

3.2. Data exchange

The interaction of the modules is carried out by means of the coupler, which receives and sends the necessary data between the system components. Below is a table of data obtained by the coupler from each of the interacting modules, indicating in which modules this data is used:

Data short name	Long name	Used in
	Atmosphere and Land Model	
zlvl	Wind velocity height	Ocean, Ice
ztlvl	Temperature and humidity height	Ocean, Ice
uatm, vatm	Wind velocity components	Ocean, Ice
potT	Potential temperature	Ocean, Ice
Tair	Air temperature	Ocean, Ice
Qa	Specific humidity	Ocean, Ice
rhoa	Air density	Ocean, Ice
Fsw	Downward shortwave radiation	Ocean, Ice
Flw	Downward longwave radiation	Ocean, Ice
Frain	Precipitation rate (rain)	Ocean
Fsnow	Precipitation rate (snow)	Ocean, Ice

cld	Cloudiness	Ocean, Ice
runoff	River discharge transport	Ocean
	Ocean Model	
sst	Sea surface temperature	Atmosphere, Ice
sss	Sea surface salinity	Ice
uocn, vocn	Components of ocean surface current velocity	Ice
tiltx, tilty	Sea surface tilt components	Ice
frzmlt	Freezing potential	Ice
	Ice Model	
ifrc	Ice area fraction	Ocean
tsfc	Temperature of the ice surface	Atmosphere
alb	Ice albedo	Atmosphere
tauxa, tauya	Wind stress	Atmosphere
lat	Latent heat flux	Atmosphere
sens	Sensible heat flux	Atmosphere
lwup	Upward longwave radiation	Atmosphere
evap	Evaporation rate	Atmosphere
Tref	Reference temperature at 10 m	Diagnostics
netsw	Net shortwave radiation	Atmosphere
melth	Heat flux into the ocean due to ice melt	Ocean
meltw	Fresh water flux into the ocean due to ice melt	Ocean
salt	Salinity flux into the ocean due to ice melt	Ocean
tauxo, tauyo	Ice-ocean stress	Ocean
swabs	Shortwave radiation passing through ice into ocean	Ocean

As a result of the coupler, the flow fields and their characteristics for different modules of the system are formed. Below is the distribution table for each module:

Data short name	Long name
	Atmosphere and Land Model
tsurf	Surface temperature
Cu	Drag-coefficient for wind stress ($\boldsymbol{\tau} = \rho C_u \mathbf{u}$)
Ct	Drag-coefficient for sensible heat flux ($H_s = \rho c_p C_T (T-T_s)$)
Ce	Drag-coefficient for evaporation flux ($E_s = \rho C_e (q-q_s)$)
albedo	Surface albedo
	Ocean Model
swave	Shortwave radiation
netheat	Sensible heat flux
prec	Precipitation rate
evap	Evaporation rate
taux, tauy	Sea-surface stress
ifrc	Ice area fraction
melt	Heat flux due to ice melt
salt	Salt flux due to ice melt
utidal, vtidal	Tidal velocity components

Ice Model	
sst	Sea surface temperature
sss	Sea surface salinity
uocn, vocn	Components of ocean surface current velocity
zlvl	Height of lower atmosphere characteristics
uatm, vatm	Wind velocity components
potT	Potential temperature
Tair	Air temperature
Qa	Specific humidity
rhoa	Air density
tiltx, tilty	Sea surface tilt
fw	Freezing potential
fsw	Downward shortwave radiation
flw	Downward longwave radiation
rain	Precipitation rate (rain)
snow	Precipitation rate (snow)

4. Results. 100-year coupled run

As a test, the model of the climate system was calculated for a period of 100 years. The initial state of the atmosphere was obtained in previous experiments with the full version of the stand-alone PlaSim model.

4.1. Ocean

Figure 2 shows the sea surface temperature obtained as a result of averaging over the last ten years of an experiment for March and September. In general, the temperature distribution is in good agreement with the known climate distributions, but it is about 2-3 degrees colder in the tropics. The results for the most critical ocean region, the Arctic, are not entirely satisfactory. The flow in the upper 100 m layer in Figure 3 corresponds to a strong cyclonic circulation mode and does not contain characteristic features of the anticyclonic circulation in the Beaufort Sea. As a result of insufficient convergence of the currents in this region, the ice field also turned out to be more dispersed than the real distribution, which led, for example, to the fact that an area north of Scandinavia and the Murmansk port were under ice (Figures 4a and b), even in summer. In summer (southern hemisphere), the ice field in the Antarctic (Figure 4c) is almost completely absent, and in winter it reaches 0.5-1 m (Figure 4d). The heat transport from the equator to the poles is somewhat overestimated in the northern hemisphere and underestimated in the southern hemisphere. However, if we consider the meridional heat transport in dynamics, it can be seen that a 100-year period is probably insufficient for its final formation, and the changes that have occurred in the last 50 years are in a positive direction (Figure 5).

IOP Conf. Series: Earth and Environmental Science **96** (2017) 012005 doi:10.1088/1755-1315/96/1/012005

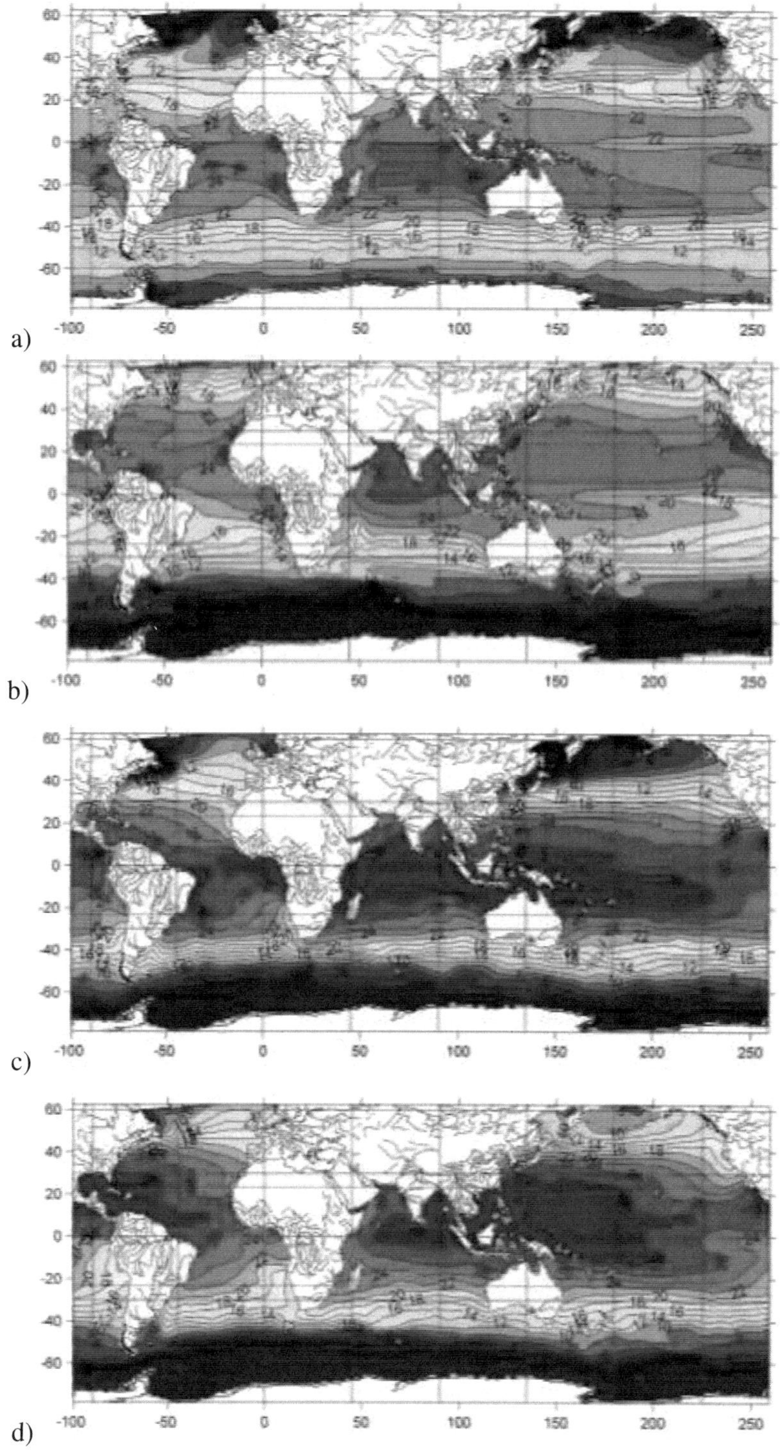

Figure 2. The sea surface temperature a) in March b) in September, averaged over the last decade of the 100-year preliminary experiment. The PHC3.0 climatological sea surface temperature c) in March d) in September [Steele et al, 2000].

IOP Conf. Series: Earth and Environmental Science **96** (2017) 012005 doi:10.1088/1755-1315/96/1/012005

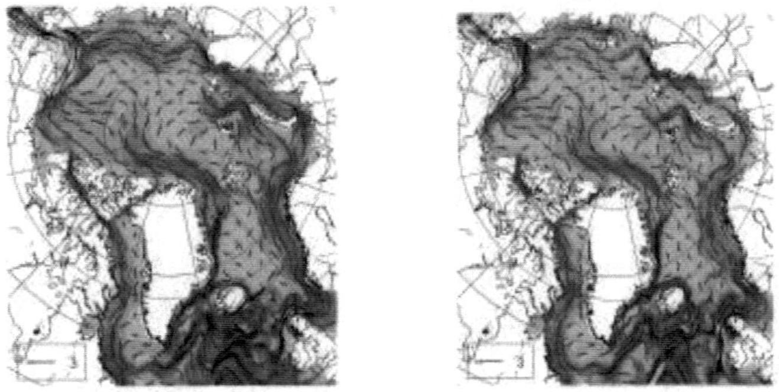

a) b)

Figure 3. The average circulation (cm/s) of the upper 100-meter layer of the ocean in the Arctic: (a) - in March, (b) - in September.

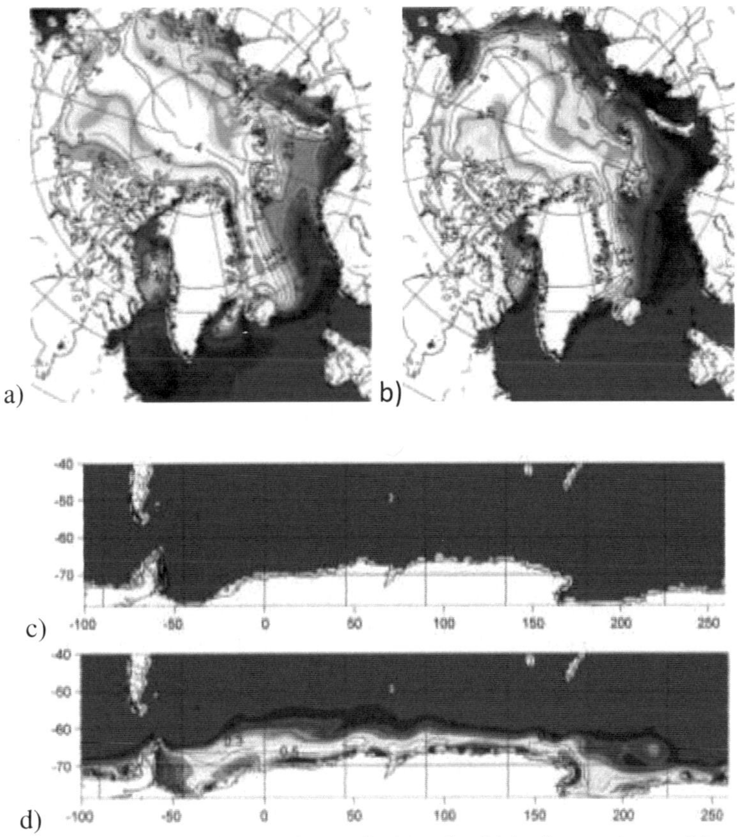

Figure 4. Ice thickness (m) in the Arctic: (a) in March, (b) in September and in the Antarctic: (c) in March, (d) in September.

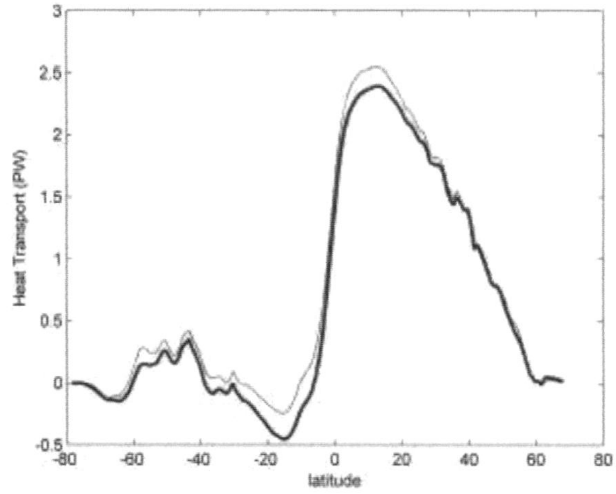

Figure 5. Meridional heat transport (PW – 10^{15} W, a positive value corresponds to a northward transport). The thin line corresponds to the averaging over the year-period 45-55, and the bold line -- over the year-period 90-100 of the preliminary 100-year experiment.

4.2. Atmosphere

Surface air temperature and precipitation fields are plotted in Figures 6-9. Figure 6 shows the surface temperature obtained as a result of averaging over the last 35 years of the experiment and over the 1979-1999 of the NCEP2 Reanalysis data for January and July.

In general, the model surface temperature distribution is in good agreement with the NCEP2 Reanalysis data, but some regions are overheated. It is about 5 degree warmer in the central part of Eurasia in January, in the southern part of this region in July, in the western part of South America in January and in North America in both months considered. In the Polar regions the model surface temperature is lower than that in the Reanalysis data in its winter seasons (in January for the North Pole and in July for the South Pole). Figure 7 shows the total precipitation obtained as a result of averaging over the last 35 years of the experiment and over the 1979-1999 of the GPCP ver. 2.3 data (www.esrl.noaa.gov) for January and July. The model data shows a good agreement with the observation data, but it is about 10-15 mm per day greater in the modeling data than in the observations in the tropics for both months considered. Some major arid regions are captured, specifically the seasonal migration of the Inter-Tropical Convergence Zone and the associated monsoon systems. The PlaSim-ICMMG-v.1.0 better correlates with the observed seasonal precipitation than the PlaSim, and it shows wetter conditions (Figure 8).

The results of modeling with the new model show that in January in the Arctic the global surface temperature is lower than the observed one: a cold bias and a comparison of the simulation results with the old model with the reanalysis data shows a warm bias (Figure 9).

The average observed January and July climatology is better simulated by the PlaSim -_ICMMG-v.1.0 model than by the PlaSim model.

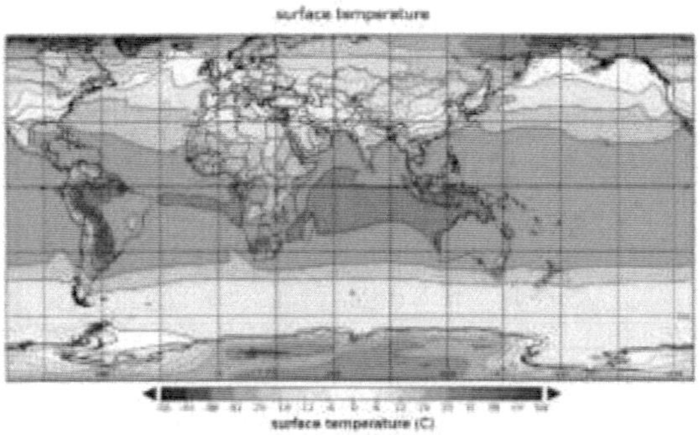

a)

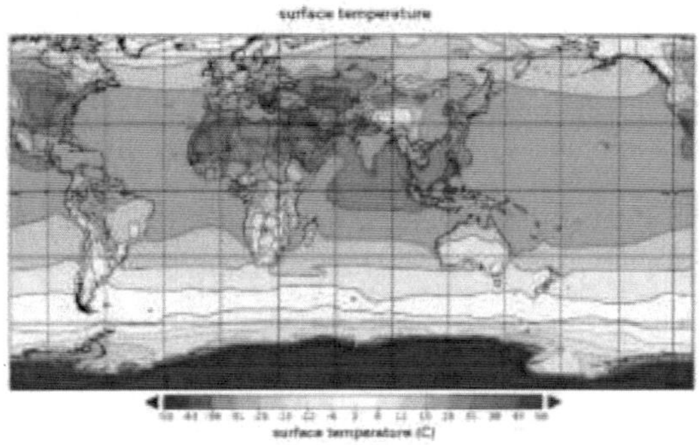

b)

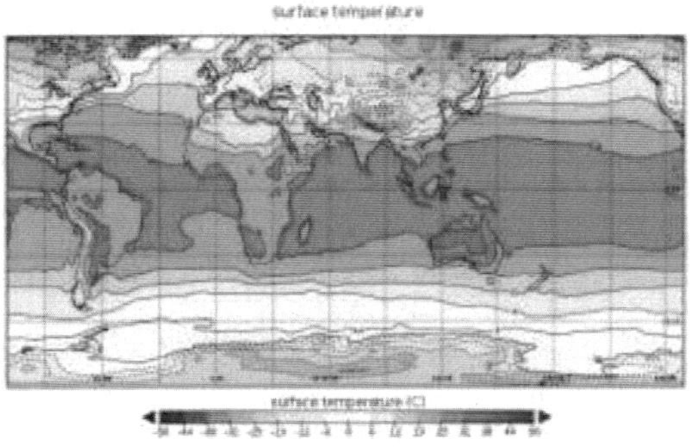

c)

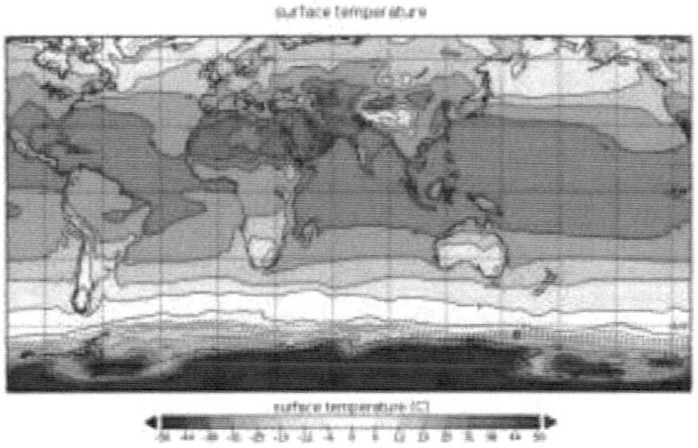

d)

Figure 6. Surface temperature averaged over the last decade of the 100-year preliminary experiment: a) in January; b) in July. The climatological surface temperature for NCEP2 Reanalysis averaged over 1979-1999 for c) January and d) July.

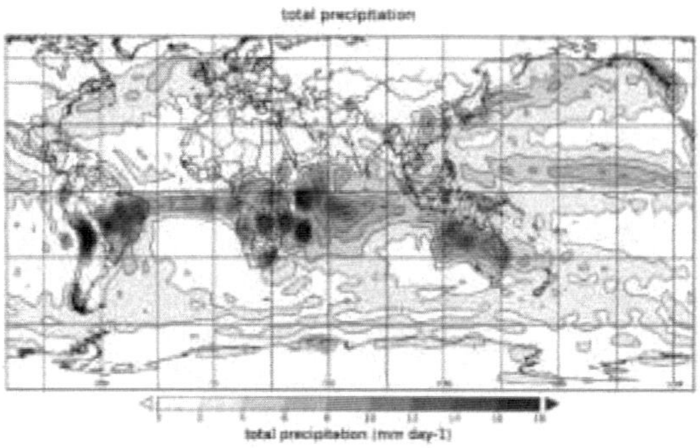

a)

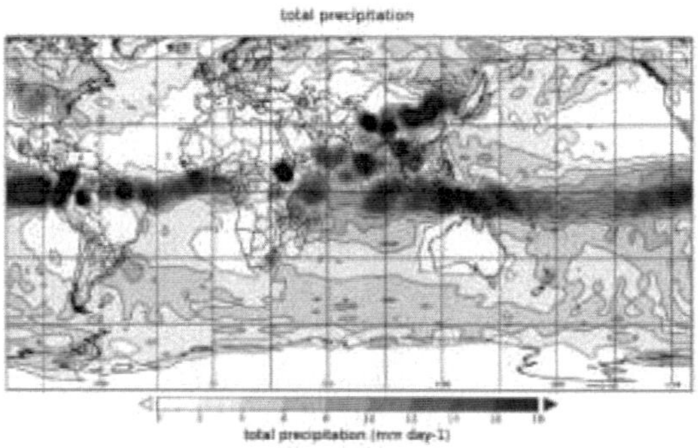

b)

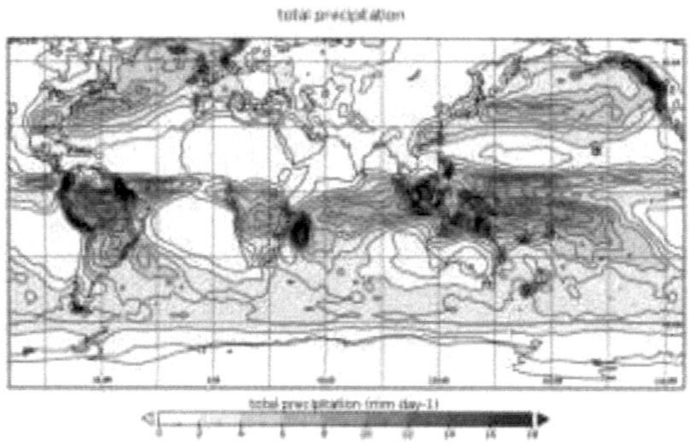

c)

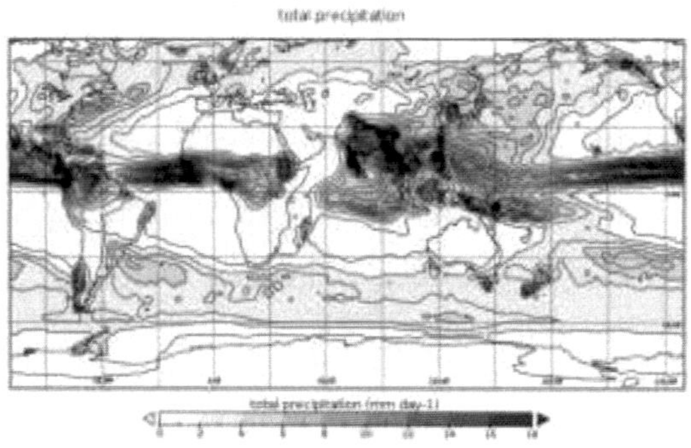

d)

Figure 7. Combined precipitation averaged over the last decade of the 100-year preliminary experiment: a) in January; b) in July. The climatological total precipitation for GPCP ver. 2.3 dataset averaged over 1979-1999 for c) January and d) July (www.esrl.noaa.gov).

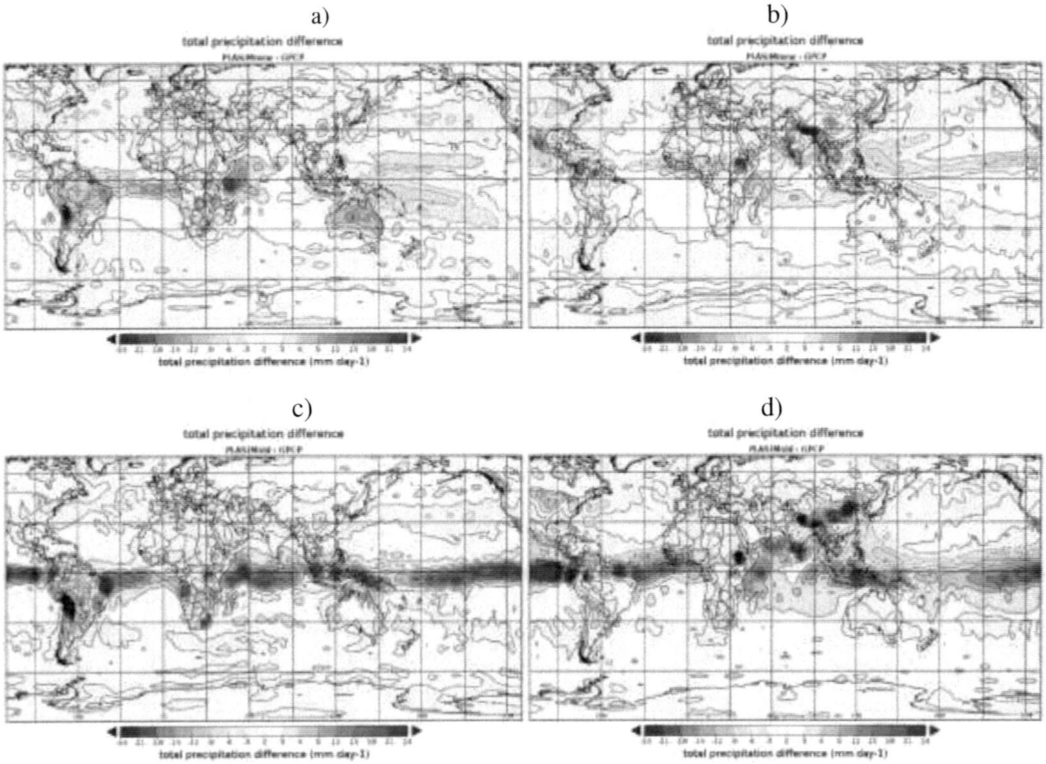

Figure 8. Differences between combined precipitation averaged over the last decade of the 100-year preliminary experiment and climatological total precipitation for GPCP ver. 2.3 dataset averaged over 1979-1999(www.esrl.noaa.gov): a) in January; b) in July for PlaSim-ICMMG-v.1.0 ; c) January and d) July for PlaSim.

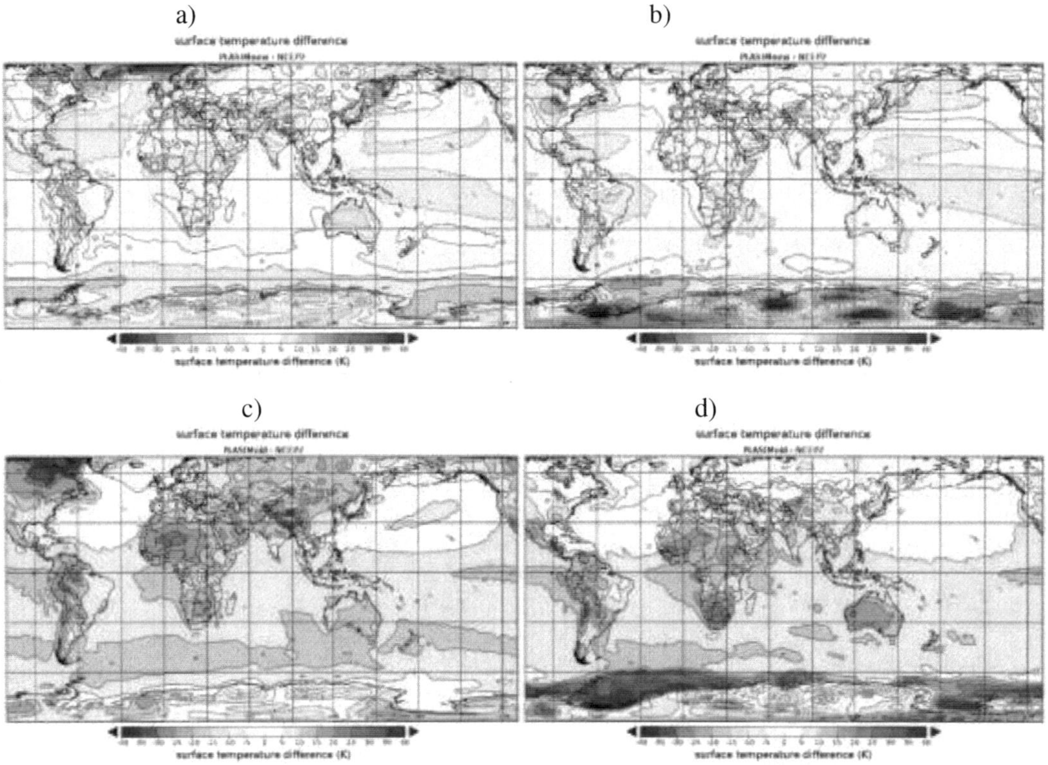

Figure 9. Seasonal surface air temperature (C) difference: (PLASIM-ICMMGv.1.0 - NCEP) - a) January, b) July; (PLASIM - NCEP) – c) January, d) July.

5. Conclusions

It is well-known that the complexity of the climate system presents a challenge to the climate theory. We try to simulate as much of climate dynamics as we can with comprehensive numerical models (for example, [33]), and we also try to understand the simplifying and highlighting key phenomena in idealized models [34].

We have presented a new intermediate complexity climate system model, PLASIM–INMMG-v.1.0, which reproduces the main parameters of the climate system. We conclude that the new model reproduces the seasonal surface temperature and precipitation reasonably well and in general better than the old version, PlaSim.

The features of the ocean circulation reproduced by the model and the resulting distribution of sea ice significantly deteriorated in comparison with the results obtained earlier [24]. The sea surface temperature in the equatorial region is 2-3 degrees below the climate data. The ice distribution is wider towards the mid-latitudes in the North Atlantic, while in the southern hemisphere the ice almost completely disappears in summer (in the southern hemisphere). Nevertheless, the dynamics of the formation of meridional heat transfer is such that it allows us to state that, firstly, a quasi-stationary state of the climate system has not yet been reached in the first 100 years of model integration. Secondly, from the point of view of heat exchange between the northern and southern hemispheres, the system develops in a positive direction, allowing us to hope that a quasi-stationary state will be achieved with climatic parameters closer to the observed ones. The achieved state, despite its difference from the observed climate, is still quite plausible, which means that the model needs some additional tuning and selection of more suitable values of the model parameters.

Acknowledgements. We gratefully acknowledge support by the RFBR (grants no. 16-05-00558 and no. 17-05-00382).

References

[1] Fraedrich K, Jansen H, Kirk E, Luksch U and Lunkeit F 2005 The Planet Simulator: Towards a user friendly model *Meteorol. Zeitschrift* **14** 299 -304
[2] Lunkeit F, Böttinger M, Fraedrich K, Jansen H, Kirk E, Kleidon A and Luksch U 2007 Planet Simulator Reference Manual Version 15.0, available at: http://epic.awi.de/29588/1/Lun2007d.pdf
[3] Petoukhov V, Ganopolski A, Brovkin V, Claussen M, Eliseev A and Kubatzki S 2000 Rahmstorf CLIMBER-2: a climate system model of intermediate complexity. Part I: model description and performance for present climate *Climate Dynamics* **16** (1) 1–17
[4] Brovkin V, Claussen M, Driesschaert E, Fichefet T, Kicklighter D, Loutre M F, Matthews H D, Ramankutty N, Schaeffer M and Sokolov A 2006 Biogeophysical effects of historical land cover changes simulated by six Earth system models of intermediate complexity *Clim. Dynam* **26** 587–600
[5] Claussen M *et al* 2002 Earth system models of intermediate complexity: closing the gap in the spectrum of climate system models *Clim. Dynam* **18** 579–86
[6] Krupchatnikov V, Kuzin V, Golubeva E, Martynova Yu, Platov G and Krylova A 2009 Hydrology and Vegetation Dynamics of the Climate System of Northern Eurasia and the Arctic Basin *Izvestiya Atmospheric and Oceanic Physics* **45** (1) 116–36
[7] Martynova Yu V, Krupchatnikov V N 2010 A study of the sensitivity of the surface temperature in Eurasia in winter to snow-cover anomalies: The role of the stratosphere *Izvestiya Atmospheric and Oceanic Physics* **46** (6) 757–69
[8] Martynova Yu V, Krupchatnikov V N 2015 Peculiarities of the Dynamics of the general atmospheric circulation in conditions of the global climate change // *Izvestiya Atmospheric and Oceanic Physics* **51** (3) 299–310
[9] Borovko I V and Krupchatnikov V N 2015 Responses of the Hadley cell and extratropical troposphere stratification to climate changes simulated with a relatively simple general circulation model *Numerical Analysis and Applications* **8** (1) 23–34
[10] Eliseev V, Coumou D, Chernokulsky A V, Petoukhov V and Petri S 2013 Scheme for calculation of multi-layer cloudiness and precipitation for climate models of intermediate *Geosci. Model Dev* **6** 1745–65
[11] Comiso J. C, Parkinson C L, Gersten R and Stock L 2008 Accelerated decline in the Arctic sea ice cover *Geophys. Res. Lett.* **35** L01703
[12] Bekryaev R V, Polyakov I V and Alexeev V A 2010 Role of polar amplification in long-term surface air temperature variation and modern Arctic warming *J. Climate* **23** 3888-906
[13] Comiso J C 2012 Large decadal decline of the Arctic multiyear ice cover *J. Climate* **25** 1176–93
[14] Screen J A and Simmonds I 2010 The central role of diminishing sea ice in recent Arctic temperature amplification *Nature* **464** 1334–37
[15] Cohen J *et al* 2014 Recent Arctic amplification and extreme mid-latitude weather *Nat. Geosci.* **7** 627–37
[16] Jaiser R, Dethloff K, Handorf D, Rinke A and Cohen J 2012 Impact of sea ice cover changes on the Northern Hemisphere atmospheric winter circulation *Tellus* **64**A 11595
[17] Francis J A and Vavrus S J 2012 Evidence linking Arctic amplification to extreme weather in mid-latitudes *Geophys. Res. Lett.* **39** L06801 doi:10.1029/2012GL051000
[18] Francis J A and Vavrus S J 2015 Evidence for a wavier jet stream in response to rapid Arctic warming *Environ. Res. Lett.* **10** 014005 doi:10.1088/1748-9326/10/1/014005
[19] Overland J, Francis J A, Hall R, Hanna E, Kim S J and Vihma T 2015 The melting Arctic and midlatitude weather patterns: are they connected? *J Clim* **28** 7917–32

[20] Shepherd T G 2016 Effects of a warming Arctic *Science* **353** 989–90
[21] Overland J E, Dethloff K, Francis J A, Hall R J, Hanna E, Kim S-J, Screen J A, Shepherd T G, Vihma T 2016 Nonlinear response of midlatitude weather to the changing Arctic *Nat Clim Change* **6** 992–99
[22] Petoukhov V, Rahmstorf S, Petri S and Schellnhuber H J 2013 Quasiresonant amplification of planetary waves and recent Northern Hemisphere weather extremes *ProcNatl Acad Sci USA* **110** 5336–41
[23] Golubeva E N and Platov G A 2009 Numerical modeling of the Arctic Ocean ice system response to variations in the atmospheric circulation from 1948 to 2007 *Izvestiya Atmospheric and Oceanic Physics* **45** (1) 137–51
[24] Kuzin V I 1985 *The finite element method in the modeling of oceanic processes* (Novosibirsk: Computing Center of SB AN SSSR) p 189
[25] Golubeva E N, Ivanov U A, Kuzin V I and Platov G 1992 A numerical modeling of the world ocean circulation with allowance for the upper quasi-homogeneous layer *Oceanologia* **32** (3) 395–405
[26] Golubeva E N and Platov G A 2007 On improving the simulation of Atlantic water circulation in the Arctic Ocean *J. Geophys. Res.* **112** doi: 10.1029/2006JC003734
[27] Leonard B P 1979 A stable and accurate convective modeling procedure based on quadratic upstream interpolation *Comput. Methods Appl. Mech. Engrg.* **19** 59–98
[28] Hunke E C and Dukowicz J K 1997 An elastic-viscous-plastic model for sea ice dynamics *J. Phys. Oceanogr.* **27** 1849–67
[29] Hibler W D 1979 A dynamic thermodynamic sea ice model *J. Phys. Oceanogr.* **9**(4) 815–46
[30] Steele M, Morley R and Ermold W 2000 PHC: A global hydrography with a high quality Arctic Ocean *J. Climate* **14** 2079-87
[31] Murray R J 1996 Explicit generation of orthogonal grids for ocean models *J. Comput. Phys.* **126** 251–73
[32] Volodin E M, Diansky N A and Gusev A V 2013 Simulation and prediction of climate changes in the 19th to 21st centuries with the Institute of Numerical Mathematics, Russian Academy of Sciences, model of the Earth climate system *Izvestiya. Atmospheric and Oceanic Physics* **49** (4) 347–66
[33] Held I 2005 The Gap between Simulation and Understanding in Climate Modeling *Bulletin of the American Meteorological Society* **86** 1609–14

A comparison of the structure and dynamics of Global atmospheric oscillation in reality and in the CMIP5 climate models

I V Serykh

Shirshov Institute of Oceanology, Russian Academy of Sciences,
36, Nahimovskiy prospekt, Moscow, Russia, 117997

E-mail: iserykh@ocean.ru

Abstract. In this paper, global fields of mean anomalies of the sea-level atmospheric pressure and the surface air temperature are constructed for positive and negative phases of a recently discovered by Russian scientists Global atmospheric oscillation (GAO) with El Niño and La Niña as its elements. This is done using observations and their re-analyses, as well as the results of experiments with climate models. A GAO index is proposed, and its spectra, as well as spectra of El Niño – Southern Oscillation (ENSO), are calculated. It is shown that some of the modern models of ocean-atmosphere general circulation of the international Coupled Model Intercomparison Project Phase 5 (CMIP5) reproduce well the GAO spatial structure. The model temporal energy spectra, however, differ from the real ones in the general variation of the energy spectra of interannual-decadal oscillations, as well as in the periods of specific peaks in this range. A comparison of CMIP5 experiments called Historical and piControl shows that the climate models with the 11-year solar activity cycle forcing reproduce the GAO periodicity more accurately. It is concluded that the differences between the model spectra and the real ones cause major errors in the predictions of the El Niño onset for more than half a year.

1. Introduction

El Niño – Southern Oscillation (ENSO) is a well-known phenomenon observed in the ocean-atmosphere system. ENSO develops in a narrow equatorial Pacific Ocean band. That is, formally it is a regional event. However, it is commonly accepted that ENSO affects the meteorological processes on interannual to decadal time scales on the Earth. Numerous publications in Russian and other languages (see, for instance, [1, 2, 3]) are devoted to this problem. Note that a dozen of papers annually consider the importance of global influence of ENSO.

Some years ago a team of researchers from the Shirshov Institute of Oceanology RAS [4] suggested that there exists a Global atmospheric oscillation (GAO) with El Niño and La Niña developing at its extreme phases. In other words, they proposed that there is a primary interaction of some hydrometeorological processes taking place on the Earth, and ENSO is a major manifestation of these processes.

To detect the GAO, the global sea-level atmospheric pressure (SLP) fields corresponding to the ten strongest El Niño and La Niña events that took place in 1950-2010 were compared. They were used as a basis for constructing a mean field of differences between the El Niño and La Niña SLP fields. A major peculiarity of the mean field of differences was the presence of a wide positive SLP region

Content from this work may be used under the terms of the Creative Commons Attribution 3.0 licence. Any further distribution of this work must maintain attribution to the author(s) and the title of the work, journal citation and DOI.
Published under licence by IOP Publishing Ltd

stretching in the zonal direction along the equator from Indonesia through the Indian Ocean, Africa, and the Atlantic to the Brazil coast. In the meridional direction this region is bounded by the 30° north and south latitudes. This region is surrounded on all sides by narrow bands of negative SLP differences. They merge and enhance in magnitude at the center and the eastern part of the Pacific Ocean crossing in the canonical zone of El Niño and La Niña development. Near the polar zones of the both hemispheres the SLP differences become positive again. To the south-west of South America there is a most pronounced area of positive SLP values. Here some researchers [5, 6] had discovered a high-power anticyclone developing simultaneously with El Niño.

Since the number of El Niño and La Niña events used to construct the mean field of differences by [4] was not large, it was not clear whether the mean field of SLP differences was statistically significant and, hence, whether the GAO was real. Therefore, in [7] the mean global field of SLP differences between the El Niño and La Niña events was recalculated with a larger number (23 and 25) of events that took place in 1920-2012. It is important that the global fields of root-mean-square SLP deviations from the mean fields of El Niño and La Niña were calculated separately for both groups of events. With these root-mean-square deviations, Student's t-test values were calculated for each geographical grid node. These values showed that the areas of positive and negative SLP differences in the GAO spatial structure are statistically highly significant (95% and even more). This proved that the GAO really exists.

This brings up the question, whether the modern coupled models of the ocean-atmosphere general circulation, that can simulate well many features of interannual variability in the Pacific Ocean tropics [8], can simulate the GAO. A comparison of the temporal energy spectra of the ENSO indices calculated with the observational data and the spectra obtained with the climate models of Coupled Model Intercomparison Project Phase 5 (CMIP5) has revealed a large scatter in the results [9, 10, 11, 12, 13]. The global and regional structures of anomalies of hydrometeorological fields at the El Niño events obtained with the CMIP5 models have also shown a wide spread in the observed anomalies [14, 15, 16, 17].

In the present paper, the global structures of SLP and the surface air temperature (SAT) differences between the El Niño and La Niña events, that is, between the positive and negative GAO phases, as well as the temporal energy spectra of the ENSO and GAO indices calculated with the observations and their re-analyses, are compared with the corresponding characteristics obtained by calculating the CMIP5 models in order to reveal possible shortcomings of the modern climate models.

2. Materials and methods of the study

The global monthly mean SLP fields used in the analysis were taken from the following sources: NCEP/NCAR Reanalysis on a 2.5°x2.5° grid over 1948-2014 [18], Met Office Hadley Center HadSLP2 on a 5°x5° grid over 1850-2014 [19], NOAA-CIRES 20th Century Global Reanalysis Version 2c on a 2°x2° grid over 1851-2011 [20], ECMWF ERA-20C on a 1°x1° grid over 1900-2010 [21], and JMA JRA-55 on a 1.25°x1.25° grid over 1958-2013 [22].

The global monthly mean values of sea surface temperature (SST) were taken from the following sources: Met Office Hadley Center HadISST on a 1°x1° grid over 1870-2014 [23], JMA COBE SST2 on a 1°x1° grid over 1850-2014 [24], and NOAA ERSST V4 on a 2°x2° grid over 1854-2015 [25, 26]. The global monthly mean data of the near-surface temperature of Met Office Hadley Center HadCRUT.4.4 on a 5°x5° grid over 1850-2014 [27] and the global monthly mean data of SAT from the above-mentioned re-analyses (NCEP/NCAR, 20thC_ReanV2c, ERA-20C and JRA-55) have also been analyzed.

For each of the above data sets at each grid node, the average annual variation in the corresponding time period was calculated, and then subtracted from the initial data. The thus obtained anomaly fields were used to calculate three indices: first, an El Niño index called the Extended Oceanic Nino Index (EONI). It is the mean SST anomaly over (5°S-5°N, 170°-80°W). There are typically more shipborne SST observations in this region than in the region (5°S-5°N, 170°-120°W) used to estimate a Niño 3.4 index that is the standard for determining El Niño. Second, the Equatorial Southern Oscillation Index

(ESOI) determined by the difference of the mean SLP anomalies between the regions of (5°S-5°N, 90°-140°E) and (5°S-5°N, 130°-80°W). Third, the Global Atmospheric Oscillation Index (GAOI), which is the next combination of normalized mean SLP anomalies over 10 areas of the maximal statistical significance of the differences between El Niño and La Niña related SLP-distributions [7]: (5°S-5°N, 35°-25°W) + (5°S-5°N, 55°-65°E) + (55°-65°N, 95°-85°W) + (65°-55°S, 95°-85°W) + (5°S-5°N, 145°-155°E) – (45°-55°N, 175°-165°W) – (45°-55°N, 15°-5°W) – (55°-45°S, 15°-5°W) – (55°-45°S, 175°-165°W) – (5°S-5°N, 95°-85°W). Unlike other ENSO and GAO indexes, GAOI contains the anomalies of the middle- and high-latitude regions, where the probability of GAO anomalies is large. Therefore, GAOI characterizes the whole planetary structure of the GAO, rather not only its tropical part. The GAOI varies synchronously with the EONI and ESOI, because its variations incorporate the Southern Oscillation. This shows that all atmospheric processes resulting in GAO evolve more or less synchronously. The linear trends in the thus obtained series of all the indices were removed by the least squares method to suppress the interdecadal climate changes.

El Niño and La Niña events were determined using EONI as follows: if this index within 5 successive months or longer exceeded +0.5°C, it was an El Niño, and if it was less than -0.5°C, it was a La Niña. The global fields of mean SLP and SAT anomalies were calculated for each event chosen in this way. These fields were averaged for El Niño and La Niña events separately. Then the difference between the thus obtained mean fields was calculated. The field of root-mean-square deviations of individual events from the mean field was determined for each type of events (El Niño and La Niña). This was used for estimating the fields of statistical significance of the differences between the mean fields according to Student's t-test with various probabilities. The positive and negative GAO phases were determined with GAOI, similarly to determining the El Niño and La Niña events. However, the root-mean-square deviation and the time in which the autocorrelation function of the GAOI reaches near-zero value served as a criterion for selection. For the GAO phases, global fields of mean SAT and SLP differences have also been constructed and the statistical significance of these anomalies has been estimated, which are very close to the fields obtained by EONI.

The temporal energy spectra of the indices were calculated using the fast Fourier transform (FFT). To provide maximum resolution of these spectra in the periods, a large group of spectra has been calculated. For this, the parts corresponding to the "windows" with widths ranging from the full length of the initial series to its half were taken, with the "windows" sliding along the entire length of the initial series. All thus obtained spectra were averaged, and the averaged spectrum was smoothed using a moving average [28].

The spectra of the so-called mappings of the indices on the annual time scale were calculated similarly. For this, a sequence of values of a month, for instance, the sequence of Januaries, Februaries, etc. was taken from the series under investigation. The "window" technique described above was used to calculate the spectrum for each sequence. Then all twelve spectra were averaged. This calculation method decreased the effect of the annual cycle on the estimation of the statistical significance of spectral density peaks, since the spectra of the mappings are not as "red" as those of typical continuous series. In fact, the range of periods under consideration in the spectra of mappings is limited to two years. However, this is sufficient for studying the GAO dynamics in 2- to 7-year periods, which are typical for El Niño.

The global fields of the mean amplitude of SLP and SAT oscillations during GAO and the energy spectra of the indices calculated using measurements and re-analyses were compared with the corresponding fields and spectra obtained similarly using the results of an experiment called Historical over 1850-2005 for the 47 coupled ocean-atmosphere general circulation models available in CMIP5: ACCESS1.0, ACCESS1-3, bcc-csm1-1, bcc-csm1-1-m, BNU-ESM, CanESM2, CCSM4, CESM1-BGC, CESM1-CAM5, CESM1-FASTCHEM, CESM1-WACCM, CMCC-CESM, CMCC-CM, CMCC-CMS, CNRM-CM5, CNRM-CM5-2, CSIRO-Mk3.6, EC-EARTH, FGOALS-s2, FIO-ESM, GFDL-CM2p1, GFDL-CM3, GFDL-ESM2G, GFDL-ESM2M, GISS-E2-H, GISS-E2-H-CC, GISS-E2-R, GISS-E2-R-CC, HadCM3, HadGEM2-AO, HadGEM2-CC, HadGEM2-ES, INM-CM4, IPSL-CM5A-LR, IPSL-CM5A-MR, IPSL-CM5B-LR, MIROC4h, MIROC5, MIROC-ESM, MIROC-ESM-

CHEM, MPI-ESM-LR, MPI-ESM-MR, MPI-ESM-P, MRI-CGCM3, MRI-ESM1, NorESM1-M, and NorESM1-ME [29]. A visual comparison of the obtained fields was performed. In addition, the differences between the fields after their linear interpolation to a 1°x1° grid were calculated. From the 47 CMIP5 models investigated, those that best reproduce the GAO global structure and the energy spectra of the indices were chosen. The method described above was used to calculate the fields and spectra obtained in an experiment called piControl for these climate models. This experiment, in contrast to Historical, has no external forcing of solar activity variations. The results of Historical and piControl experiments were compared to estimate the influence of external forces on the GAO and ENSO periodicity.

3. Discussion of the results

The comparison of the global fields of SLP and SAT differences between the GAO opposite phases constructed using observations and re-analyses has shown that they are very similar. Therefore, the present paper provides only for these fields an example of NOAA-CIRES 20th Century Reanalysis (figures 1a and 2a).

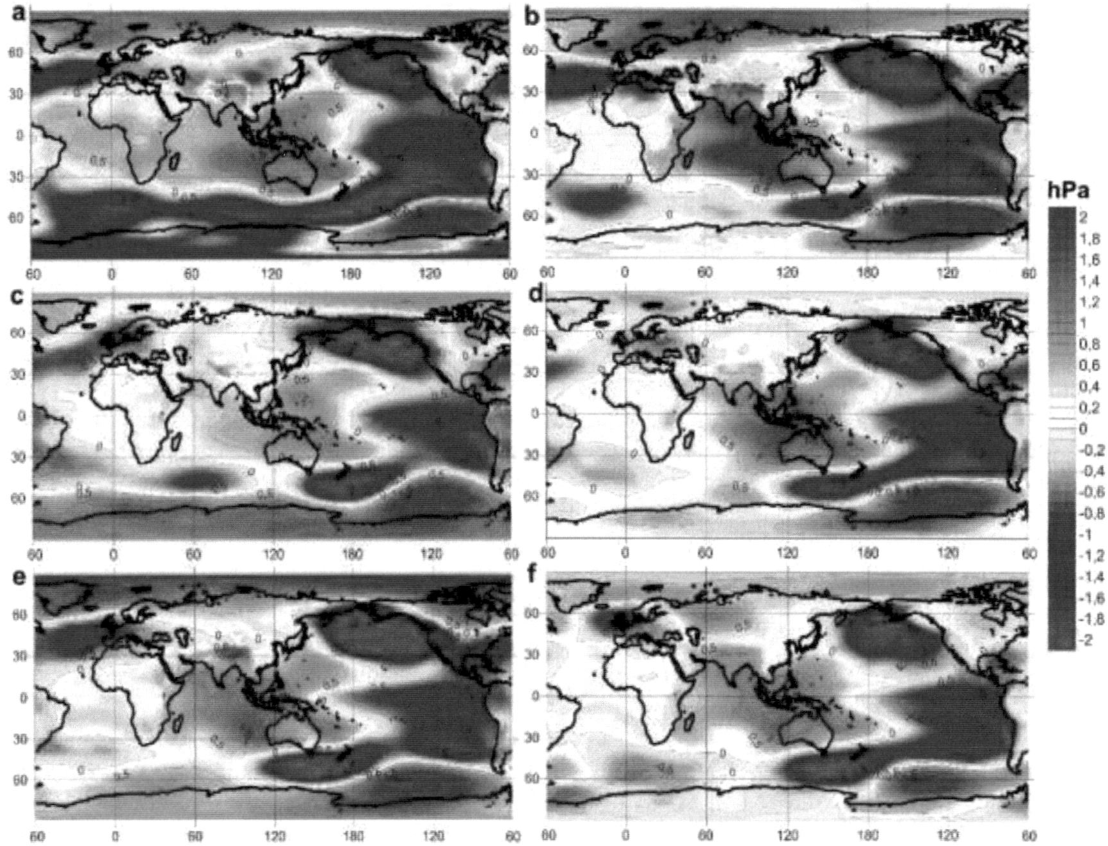

Figure 1. Fields of mean differences of sea-level pressure (SLP) between El Niño and La Niña events (positive and negative GAO phases) for NOAA-CIRES 20th Century Reanalysis v2c (a) and CMIP5 Historical experiment with climate models: CESM1-CAM5 (b), CMCC-CM (c), CNRM-CM5 (d), GFDL-CM3 (e) and HadGEM2-ES (f).

The SLP field of GAO (figure 1a) has a planetary structure with some symmetry about the equator and the meridian of 90°W, according to the location and relief of the continents. There is a wide zone of positive SLP GAO values in the tropics and subtropics of the Atlantic and Indian Oceans and the western part of the Pacific Ocean surrounded by bands of negative values that intersect in the eastern part of the Pacific Ocean (120°-90°W) near the equator. Two zones of positive SLP anomalies are located almost symmetrically in high latitudes to the north and south of this intersection.

The above-described structure of the SLP oscillations caused by GAO is in good agreement with the accompanying SAT changes (figure 2a). They reveal a global structure of the temperature oscillations symmetric (with a correction for the continents) about the equator and the meridian passing through 120°W. There is a wide zone of positive SAT values in the tropics and subtropics of the Atlantic and Indian Oceans surrounded by bands of negative values. However, at their point of intersection in the central-eastern part of the Pacific Ocean tropics there is a "tongue" of positive SAT anomalies that is typical for El Niño. To the north and south of the "tongue", in high latitudes, there are two zones of positive values of SAT anomalies located almost symmetrically.

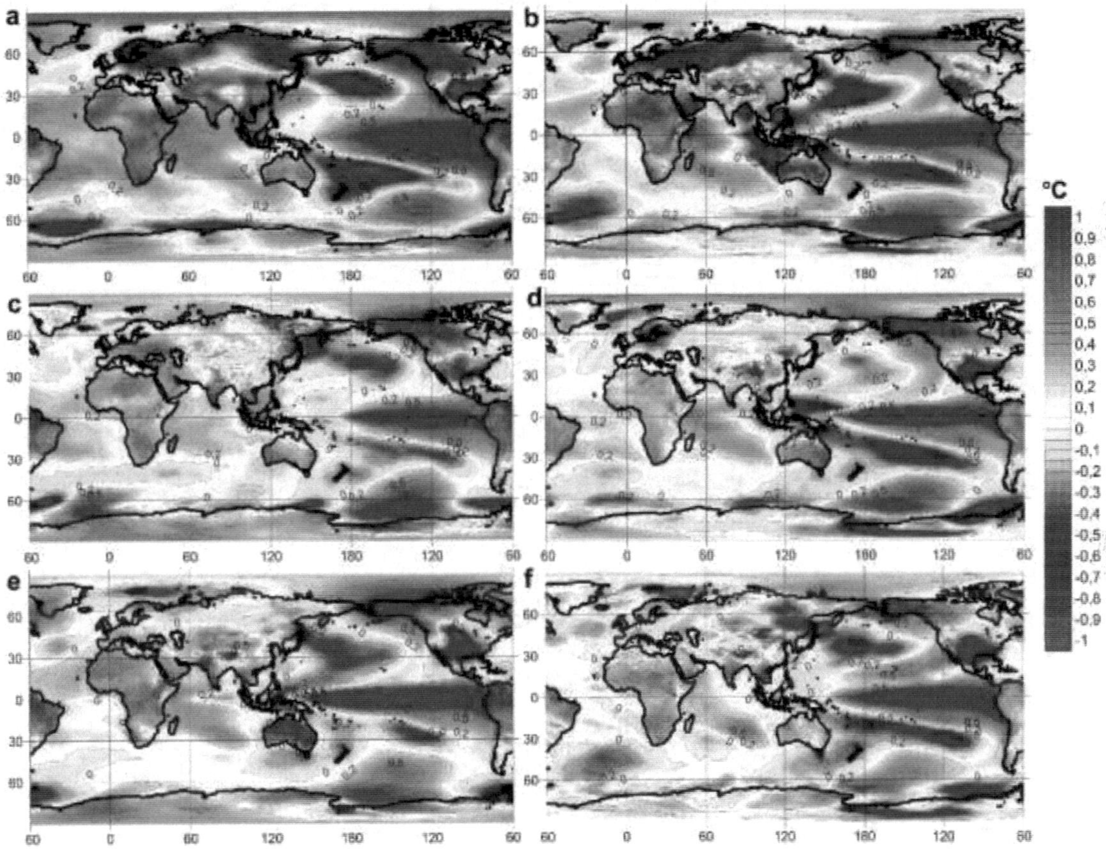

Figure 2. Fields of mean differences of surface air temperature (SAT) between El Niño and La Niña events (positive and negative GAO phases) for NOAA-CIRES 20th Century Reanalysis v2c (a) and CMIP5 Historical experiment with climate models: CESM1-CAM5 (b), CMCC-CM (c), CNRM-CM5 (d), GFDL-CM3 (e) and HadGEM2-ES (f).

The fields of SLP and SAT differences between the opposite GAO phases calculated with the results of the CMIP5 Historical experiment of climate models that are in the best agreement with the observations and their re-analyses are presented in figures 1b-1f and 2b-2f. In the SLP fields obtained by the models, the differences are underestimated (in comparison to the observations) in the tropics of the Atlantic Ocean and Africa (figure 1). This shows that in the models the teleconnections between the Indo-Pacific Walker circulation cells and the Atlantic and African ones are weaker. In the models results the bands of SLP negative values surrounding the wide tropical zone of positive values are not clearly defined and have more breaks than the similar bands in the observations. This suggests that the interrelations between the tropics and mid-latitudes in the Hadley circulation are not fully taken into account in the models. In contrast to the observations, the SLP spatial structures obtained by the models better reproduce the corresponding SAT structures. Model's SLP spatial structures reveal the "tongue" of positive SST anomalies in the Pacific Ocean tropics typical for El Niño. This indicates that the effect of SST changes on SLP is overestimated in the models. In some models (not shown in figure 2) this SST "tongue" stretches too far to the western part of the Pacific Ocean covering the region of Indonesia, which results from the fact that equatorial upwelling in the central and western parts of the Pacific Ocean is overestimated in the models [8]. Note that the structure of SAT fields calculated by the models is less smooth over the continents than that of similar fields in the re-analyses. This indicates that the influence of relief in these models is overestimated.

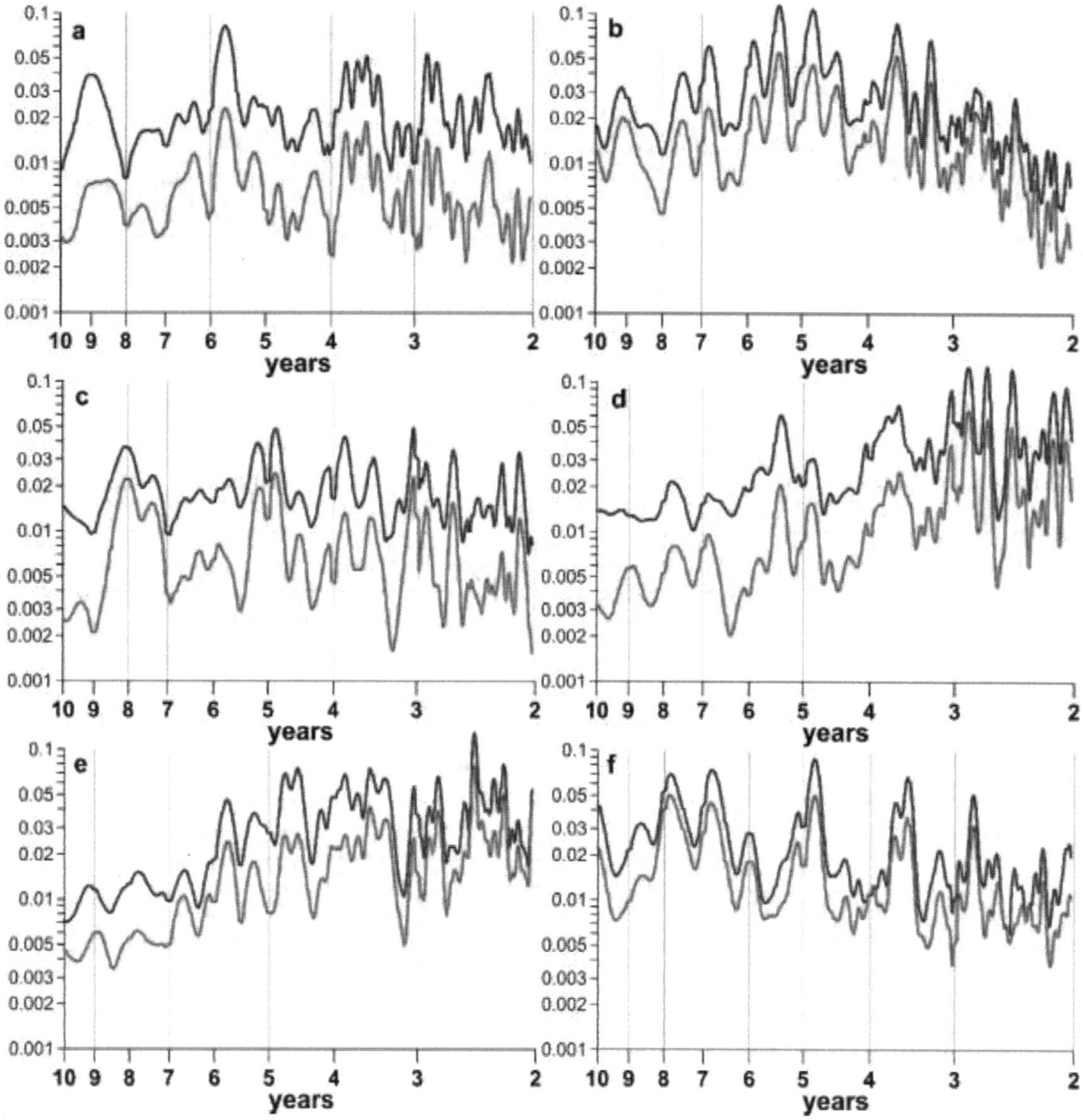

Figure 3. Energy spectra of EONI (red curves) and ESOI (blue curves) for NOAA-CIRES 20th Century Reanalysis v2c (a) and CMIP5 Historical experiment with climate models: CESM1-CAM5 (b), CMCC-CM (c), CNRM-CM5 (d), GFDL-CM3 (e) and HadGEM2-ES (f).

The best way to understand how the models reproduce ENSO and GAO temporal dynamics is to consider the observed and simulated temporal energy spectra of EONI and ESOI. In figure 3 these spectra are shown for the annual cycle mappings. These spectra allow more reliable estimation of the statistical significance of the spectral density peaks on scales from 2 to 10 years, since they are not as "red" as the spectra of their continuous analogs. It was found that the locations of peaks on the axis of the periods are very similar for all EONI, ESOI and GAOI spectra. This means that they can be analyzed simultaneously as one and the same spectrum.

Figure 3a shows that this real spectrum is practically "white", that is, its spectral density is the same if it is considered in a window of about a year or wider. From the model spectra presented, only the spectrum of the CMCC-CM model has this property (figure 3c). The spectra of two models, CESM1-CAM5 (figure 3b) and HadGEM2-ES (figure 3f), are "red". That is, the spectral density, in general, increases with the increasing period up to about 5-6 years. Then it does not increase, but even begins to slightly decrease. The spectral density of the CNRM-CM5 (figure 3d) and GFDL-CM3 (figure 3e) models, in general, decreases with the increasing period.

There are considerable differences between the real and model spectra also in the locations and powers of the spectral density peaks. It has long been accepted by most researchers that the major peaks in the ENSO spectra are statistically significant [30, 31]. This conclusion is very important, since some theoretical studies [32, 33, 34] of the periodically variable heat flux to the climate system from the Sun have shown that the appearance of peaks in the ENSO spectra in combinational harmonics of the annual period cannot be avoided.

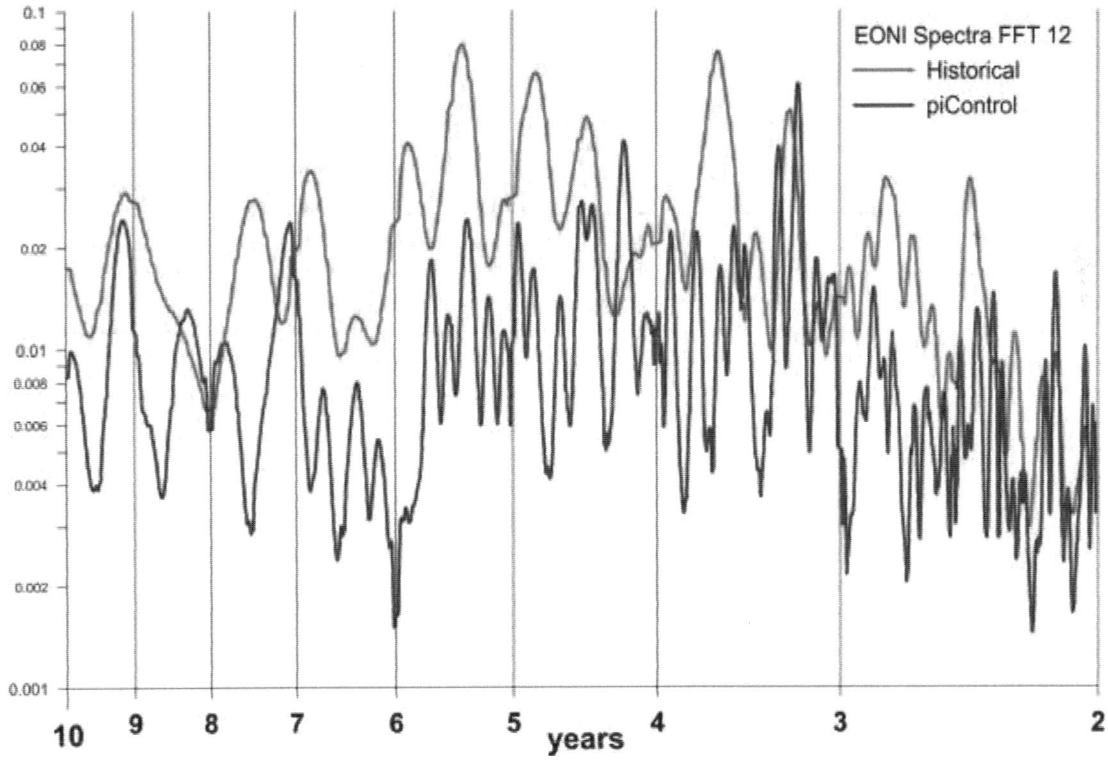

Figure 4. Energy spectra of normalized EONI of CESM1-CAM5 model in experiments Historical (red curve) and piControl (blue curve).

In the real spectrum (figure 3a) the major peaks are at periods of 5.6, 3.7 and 2.8 years. These are not combinational harmonics of the annual period. However, they can be considered as superharmonics 1:2, 1:3 and 1:4 of the 11-year solar activity cycle. This means that the climate system responds to solar activity, although the energy of this action is very small. The response of the system to this small action is nonlinear. Therefore, it is evident not in the 11-year period itself, but on its superharmonics. This hypothesis has already been made [35].

The spectra of CMIP5 models shown in figure 3b-f were calculated using the results of the Historical experiment, which has 11-year solar cycle forcing. Therefore, some of the models, for instance, CESM1-CAM5 (figure 3b) and GFDL-CM3 (figure 3e), also have peaks in periods close to the above superharmonics. However, in the spectra calculated with the results of piControl experiment (without external forcing) these peaks are almost invisible in comparison to the other more powerful peaks (figure 4). Note also that the groups of powerful peaks in the real spectrum at periods of 2.4, 3.6 and 4.8 years can be considered as subharmonics 1:2, 1:3 and 1:4 of the Chandler wobble of Earth's poles with a period of about 1.2 years [36]. It was shown in [37] that this wobble can excite a warm surface current near the Pacific coast of Central America, that is, positive SST anomalies typical for El Niño. In the CMIP5 models the Chandler wobble is not taken into account. Therefore, such peaks are absent in most of the model spectra.

4. Conclusions

The comparison of the data of re-analyses and integrations of some modern climate models has shown that, on the whole, some of the coupled ocean-atmosphere general circulation models used in the international project CMIP5 reproduce well the planetary structures of anomalies in the sea-level atmospheric pressure (SLP) and surface air temperature (SAT) fields observed during El Niño and La Niña events. These planetary structures have been recently detected in the real atmosphere and called the Global atmospheric oscillation (GAO). There certainly exist numerous quantitative differences between the elements of the real and simulated GAO spatial structures.

An index, which is a combination of the mean SLP anomalies in 10 areas of the GAO global field, has been developed to characterize the GAO temporal dynamics. The temporal energy spectra of the GAO and ENSO indices on scales from 2 to 10 years have been estimated using observations and results of integrations of the CMIP5 models. These calculations have shown that there exist differences in the real and simulated spectra.

All the spectra, both the real and simulated ones, have many statistically significant peaks of spectral density. However, their location in the axis of the periods of the model spectra differs considerably from that of the real spectra. There are three major peaks in the real spectra for periods of 5.6, 3.7 and 2.8 years, that is, superharmonics of the 11-year solar activity cycle. From the model spectra presented, only two models have peaks for periods close to those for the above-mentioned periods if these spectra are calculated using the results of a Historical experiment in which the changes in solar activity is taken into account. If the spectra are calculated with the results of these models in the piControl experiment (without external 11-year solar activity cycle forcing), no peaks are observed on the above superharmonics. This is indirect evidence that the major peaks in the real spectra can be considered as nonlinear responses of the real climate system to the 11-year solar activity cycle.

The real spectra also have a sequence of peaks for periods of 2.4, 3.6 and 4.8 years which can be interpreted as subharmonics of the Chandler wobble of the Earth's poles, whose period is about 1.2 years. In none of the CMIP5 experiments the Chandler wobble was taken into account. Therefore, it is not surprising that the most of model spectra do not reproduce such peaks.

Taking into account the differences in reproducing the GAO spatial structure by the modern climate models and, especially, the temporal spectra of the GAO and ENSO indices, it is clear why the El Niño and La Niña events cannot be presently predicted for more than half a year. It seems that to overcome this so-called "spring predictability barrier" some periodic external forces on the climate system, such as the changes in solar activity and the Chandler wobble of the Earth's poles, should be taken into account.

Acknowledgments
This work was supported by the Russian Science Foundation (project no. 14-50-00095).

References

[1] Voskresenskaya E N, Polonsky A B 1993 Air pressure fluctuations in the North Atlantic and their relationship with El Nino - southern oscillations *Physical oceanography* **4** 275–82

[2] McPhaden M J, Zebiak S E, Glantz M H 2006 ENSO as an integrating concept in Earth science *Science* **314** 1740–45

[3] Peng J B, Chen L T, Zhang Q Y 2014 The relationship between the El Nino/La Nina cycle and the transition chains of four atmospheric oscillations. Part I: The four oscillations *Adv. Atmos. Sci.* **31** 2 468–79

[4] Byshev V I, Neiman V G, Romanov Yu A, Serykh I V 2012 El Niño as a consequence of the global oscillation in the dynamics of the Earth's climatic system *Doklady Earth Sciences* **446** (1) 1089–94

[5] Peterson R G, White W B 1998 Slow teleconnections linking the Antarctic Circumpolar Wave with the tropical El Nino – Southern Oscillation *J. geophys. Res.* **103** 24573–83

[6] Romanov Y A, Romanova N A, Romanov P Y 2008 Distribution of icebergs in the Atlantic and Indian ocean Sectors of the Antarctic region and its possible links with ENSO *Geophys. Res. Lett.* **35** LO2506

[7] Byshev V I, Neiman V G, Romanov Yu A, Serykh I V, Sonechkin D M 2016 Statistical significance and climatic role of the Global Atmospheric Oscillation *Oceanology* **56** (2) 165–71

[8] Volodin E M, Diansky N A 2004 El Nino in a coupled ocean-atmosphere general circulation model *Russian Meteorology and Hydrology* **12** 5–14

[9] Bellenger H, Guilyardi E, Leloup J, Lengaigne M, Vialard J 2014 ENSO representation in climate models: from CMIP3 to CMIP5 *Clim. Dyn.* **42** 1999–2018

[10] Privalsky V, Yushkov V 2015 ENSO influence upon global temperature in nature and in CMIP5 simulations *Atmos. Sci. Let.* **16** 240–45

[11] Rao J, Ren R C 2014 Statistical characteristics of ENSO events in CMIP5 models *Atmos. Oceanic Sci. Lett.* **7** 546–52

[12] Ren H L, Zuo J, Jin F F, Stuecker M F 2015 ENSO and annual cycle interaction: the combination mode representation in CMIP5 models *Clim. Dyn.* DOI 10.1007/s00382-015-2802-z

[13] Stevenson S L 2012 Significant changes to ENSO strength and impacts in the twenty-first century: Results from CMIP5 *Geophys. Res. Lett.* **39** L17703

[14] Hurwitz M M *et al* 2014 Extra-tropical atmospheric response to ENSO in the CMIP5 models *Clim. Dyn.* DOI 10.1007/s00382-014-2110-z

[15] Jha B, Hu Z, Kumar A 2014 SST and ENSO variability and change simulated in historical experiments of CMIP5 models *Clim. Dyn.* **42** 2113–24

[16] Oh J H, Shin D W, Cocke S D, Baigorria G A 2014 ENSO Teleconnection Pattern Changes over the Southeastern United States under a Climate Change Scenario in CMIP5 Models *Advances in Meteorology* **2014** 648197

[17] Weare B C 2013 El Nino teleconnections in CMIP5 models *Clim. Dyn.* **41** pp 2165–2177

[18] Kalnay E *et al* 1996 The NCEP / NCAR 40-year reanalysis project *Bull. Amer. Meteor. Soc.* **77** 437–71

[19] Allan R J, Ansell T J 2006 A new globally-complete monthly historical gridded mean sea level pressure data set (HadSLP2): 1850-2004 *J. Climate* **19** 5816–46

[20] Compo G P, Whitaker J S, Sardeshmukh P D et al 2011 The Twentieth Century Reanalysis Project *Quarterly J. Roy. Meteorol. Soc.* **137** 1–28

[21] Stickler A *et al* 2014 ERA-CLIM: Historical Surface and Upper-Air Data for Future Reanalyses *Bull. Amer. Meteor. Soc.* **95** (9) 1419–30

[22] Kobayashi S *et al* 2015 The JRA-55 Reanalysis: General Specifications and Basic Characteristics *J. Met. Soc. Jap.* **93** (1) 5–48

[23] Rayner N A *et al* 2003 Global analyses of sea surface temperature, sea ice, and night marine air

temperature since the late nineteenth century *J. Geophys. Res.* **108** (D14) 4407

[24] Hirahara S, Ishii M, Fukuda Y 2014 Centennial-scale sea surface temperature analysis and its uncertainty *J. of Climate* **27** 57–75

[25] Huang B *et al* 2015 Extended reconstructed sea surface temperature version 4 (ERSST.v4). Part I: Upgrades and intercomparisons *J. Clim.* **28** (3) 911–30

[26] Liu W *et al* 2015 Extended reconstructed sea surface temperature version 4 (ERSST.v4): Part II. Parametric and structural uncertainty estimations *J. Clim.* **28** (3) 931–51

[27] Jones P D, Lister D H, Osborn T J 2012 Hemispheric and large-scale land surface air temperature variations: An extensive revision and an update to 2010 *J. Geophys. Res.* **117** D05127

[28] Serykh I V, Sonechkin D M 2017 Manifestations of motions of the Earth's pole in the El Niño – Southern Oscillation rhythms *Doklady Earth Sciences* **472** (2) 256–59

[29] Taylor K E, Stouffer R J, Meehl G A 2012 Overview of CMIP5 and the experiment design *Bull. Am. Meteor. Soc.* **93** 485–98

[30] Trenberth K E 1976 Spatial and temporal variations of the Southern Oscillation *Quart. J. Roy. Meteor. Soc.* **102** 639–53

[31] Torrence C, Compo G P 1997 A practical guide to wavelet analysys *Bull. Amer. Meteorol. Soc.* **79** (1) 61–78

[32] Jin F F, Neelin J D, Ghil M 1994 El Nino on the devil's: annual subharmonic steps to chaos *Science* **264** 70–72

[33] Jin F F, Neelin J D, Ghil M 1996 El Nino/Southern Oscillation and the annual cycle: Subharmonic frequency locking and aperiodicity *Physica D.* **98** 442–65

[34] Tziperman E, Stone L, Cane M A et al 1994 El Nino chaos: overlapping of resonances between the seasonal cycle and the Pacific Ocean – atmosphere oscillator *Science* **264** 72–74

[35] Vakulenko N V, Sonechkin D M 2011 Evidence of the solar Activity's effect on El Nino Southern Oscillation *Oceanology* **51** (6) 935–39

[36] Sidorenkov N S 2009 *The interaction between Earth's rotation and geophysical processes* (Weinheim: Wiley-VCH & Co. KCaA) p 305

[37] Serykh I V, Sonechkin D M 2016 Confirmation of the oceanic pole tide influence on El Niño *Sovremennye Problemy Distantsionnogo Zondirovaniya Zemli iz Kosmosa* **13** (2) 44–52

Simulation of the global ocean thermohaline circulation with an eddy-resolving INMIO model configuration

K V Ushakov[1,2] and R A Ibrayev[1,2]

[1] Institute of Numerical Mathematics, Russian Academy of Sciences, Moscow, Russia
[2] Shirshov Institute of Oceanology, Russian Academy of Sciences, Moscow, Russia

Email: ushakovkv@mail.ru

Abstract. In this paper, the first results of a simulation of the mean World Ocean thermohaline characteristics obtained by the INMIO ocean general circulation model configured with 0.1 degree resolution in a 5-year long numerical experiment following the CORE-II protocol are presented. The horizontal and zonal mean distributions of the solution bias against the WOA09 data are analyzed. The seasonal cycle of heat content at a specified site of the North Atlantic is also discussed. The simulation results demonstrate a clear improvement in the quality of representation of the upper ocean compared to the results of experiments with 0.5 and 0.25 degree model configurations. Some remaining biases of the model solution and possible ways of their overcoming are highlighted.

1. Introduction

Ocean general circulation models have been used for half a century for the numerical investigation of ocean dynamics and its variability on a wide range of spatial and temporal scales, from weather forecast to climate change assessments. During this period, the model resolution has increased by two orders of magnitude (from several degrees to hundredths of a degree), and the complexity of physical parameterizations has also greatly increased. Nowadays the development of seamless models which are able to work on various spatio-temporal scales [1] is especially important.

This work follows the line of the CORE-I [2] and CORE-II [3] projects, and continues the studies of [4,5], which discussed some qualitative characteristics of the World and Atlantic Oceans general circulation simulated by the INMIO model with a resolution of 0.5° and 0.25°. The new model configuration is based on [6] and incorporates several modernizations allowing one to perform multi-year numerical studies of ocean dynamics with a resolution of 0.1°. According to some works, this resolution is the minimum necessary requirement for reproducing ocean mesoscale eddies and dynamical structures, such as western boundary currents [7,8] and Greenland-Scotland Ridge overflows [9]. Global ocean simulations of such resolution require powerful supercomputers and scalable numerical methods. Software tools that provide an environment for such calculations were created within the framework of studies [10,11] and include, first of all, a fully parallel coupler and I/O procedures, as well as effective high-level means of interprocessor communication. The use of explicit numerical methods for the ocean dynamics equations approximation allowed us to achieve a high scalability level of the model program code with a trivial two-dimensional decomposition of the computational domain. However, it demanded greater attention towards stability issues, particularly to self-consistency of numerical schemes on the difference level. The current version of the 0.1° INMIO

Content from this work may be used under the terms of the Creative Commons Attribution 3.0 licence. Any further distribution of this work must maintain attribution to the author(s) and the title of the work, journal citation and DOI.
Published under licence by IOP Publishing Ltd

model configuration is able to stably perform calculations with no Laplacian viscous terms, with small heat and salt diffusivities and a flux-corrected transport scheme for the tracer equations.

The purpose of this work is a primary study of the eddy-resolving ability of the model to reproduce the large-scale thermohaline characteristics during the first years of calculations of the spinup of dynamic processes in the upper layers of the model ocean. For this, a 5-year numerical experiment has been performed, and the results of the last year have been analyzed. The main method of analysis consists in a comparison of the annual mean model temperature and salinity fields with the WOA09 climatology data [12,13].

2. The World Ocean model and experiment setup

The INMIO model implements the three-dimensional primitive equations system of ocean dynamics with the Boussinesq and hydrostatic approximations by means of a finite-volume method on an arbitrary locally-orthogonal B-type grid and in vertical z-coordinates. The atmosphere-ocean interface is described by a nonlinear free-surface condition with explicit formulation of heat, momentum, and water fluxes calculated by the CORE boundary layer bulk formulae [14]. The lateral boundaries impose free slip and zero fluxes conditions, and the bottom friction is parameterized according to a quadratic drag law. The sea ice is described by a thermodynamic model [15].

In this study, the momentum advection is calculated by a centered difference scheme and the tracer advection utilizes a scheme of flux-corrected transport [16]. The time derivatives are approximated by the leap-frog scheme with a periodic addition of an Euler step for contraction of the two numerical modes. For efficient use of computational resources, the fast barotropic dynamics is modeled separately from the baroclinic processes by means of a shallow-water-type system of equations [17] solved with a small time step and a numerical scheme utilizing a massively-parallel algorithm with overlaps [18]. To provide numerical stability, the momentum and tracer transport equations are appended by biharmonic filters. The vertical mixing is parameterized by the Munk-Anderson scheme with activation of convective adjustment in case of an unstably stratified water column.

The experiment setup of this work corresponds to the atmospheric boundary conditions of the CORE-II protocol for 1978-1982. A 3600×1800 horizontal grid is constructed according to a tripolar scheme [19] with poles in Siberia, North-Western Canada, and on the true South Pole, and has a resolution of 0.1° in the spherical part south of 60°N. The horizontal mixing coefficients are: zero for Laplacian viscosity, -18.e9 m^4/s for biharmonic viscosity on the equator with scaling proportional to the third power of the gridbox area square root, 100 m^2/s for the tracer diffusion on the equator with a linear scaling to the gridbox area square root. The shallow water equations viscosity and biharmonic tracer diffusivity are not used in this experiment. The background values of vertical viscosity and diffusivity are 10^{-4} and 10^{-5} m^2/s, and the maximal values in places with small Richardson numbers are 10^{-2} and 10^{-3} m^2/s, respectively. The time step is 4 min. for the baroclinic dynamics equations and 5 sec. for the shallow water equations of the barotropic processes.

The monthly mean continental runoff is defined as the precipitation spread over the ocean surface close to the coasts and river estuaries. To avoid a model sea level drift, a surface water flux normalization condition is applied, i.e. the global mean value is subtracted from the total water flux, *Prec–Evap+Runoff*. The initial temperature and salinity conditions are taken from the annual mean WOA09 data, the ocean current velocities and sea ice cover are initialized as zero. The ocean bottom topography is interpolated from the ETOPO5 database [20] excluding the internal continental water reservoirs and the Black Sea. The sea surface temperature or salinity relaxation is not used.

3. Results of the numerical experiment

The distributions of ocean temperature and salinity can characterize the internal model dynamics and can be used to assess the quality of the simulated model currents, the vertical mixing processes, and the formation of water masses. Since the state of the ocean surface is closely related to the atmospheric forcing and experiences strong interannual variation, and the deep model layers require longer periods of calculation for the spinup of dynamics, we consider for the initial analysis in Figure 1 the fields of annual mean temperature and salinity biases against the WOA09 data averaged over the upper 700-metre layer of the ocean, as done in [3,5].

Since the observational data used in the WOA09 climatology are generally smoothed, their use for verification of model solutions will be justified only on sufficiently large spatial scales. First of all, the INMIO model generally slightly heats the upper ocean layers, which is typical for the same-type z-coordinate models [2,4] and may be associated with parameterization errors of vertical mixing, in particular, with the viscosity of the transport scheme used. The positive temperature bias in the East Equatorial Pacific shows that the model reproduces well the strong El-Niño event of 1982-1983.

The intense jet streams of the Gulf Stream and Kuroshio are poorly resolved by the WOA09 data and, therefore, in the model solution bias map they are manifested in the form of narrow coastal cold tongues, until their separation from the continental slope. However, on the further path of these currents, quasi-zonally oriented biases of temperature and salinity, negative to the South and positive to the North, reveal inaccurate reproduction of the currents' routes with typical deviation to the North. However, in comparison with an Atlantic simulation [5] of 0.25° resolution, the rate of warm and salty water deviation to the North has declined. On the contrary, the dipole temperature and salinity biases in the mid-latitudes of the Central Atlantic and the positive bias to the east of Iceland have not weakened with increased resolution compared to [5]. In [21], no relationship of these biases with errors of deep vertical mixing was found and, hence, we can assume that their occurrence is due to inaccurate modeling of the atmosphere–ocean interaction or atmospheric forcing errors of the CORE-II data, both causing errors in the simulated currents' pathways.

The positive temperature and salinity biases off the western coast of South Africa are apparently due to the poor upwelling resolution caused by some errors of atmospheric forcing representation. The Antarctic Circumpolar and Agulhas currents, as well as the Falkland and Brazil currents confluence region, are places of intense mesoscale activity. Its quasi-stationary elements sustained by the model but not resolved in WOA09, such as jets and standing eddies, are clearly manifested on the temperature and salinity bias maps. Finally, the extensive salinity biases near the shores of some seas, especially in the Arctic, may require an improvement in the shelf vertical mixing and fresh water distribution parameterizations.

The vertical structure of the concerned temperature and salinity biases is presented in Figures 2 and 3, where their zonal mean distributions with latitude and depth are shown for the entire World Ocean and its basins reflecting the properties of large-scale water mass ventilation. In most cases, especially for salinity, the anomalies reach depths of about 500 m, indicating a characteristic depth scale of the layer that had spun up during the time of the model experiment. Deep-lying biases of heat and salt distribution in the 60-70°N latitudes of the Atlantic Ocean may emerge from the overestimated intensity of deep convection, seen also in [21] and possibly caused by errors of surface temperature and salinity fields or incorrect representation of mixing in case of steep isopicnals [2]. On the contrary, the dipole bias in the North Indian Ocean may be caused by insufficient intensity of mixing.

The increased temperature on the latitudes of the Antarctic Intermediate Water in the Southern Ocean may be a result of the excessive mixed layer depths near the regions of Subantarctic Mode Water formation or incorrect subduction and isopicnal mixing of water masses in the area [2]. Typical maximum temperature biases are reached in the equatorial regions at a depth of 100-200 m. They appear as two cores with a diffused thermocline near the equator in the Atlantic, with a shift to the north in the Pacific and to the south in the Indian Ocean.

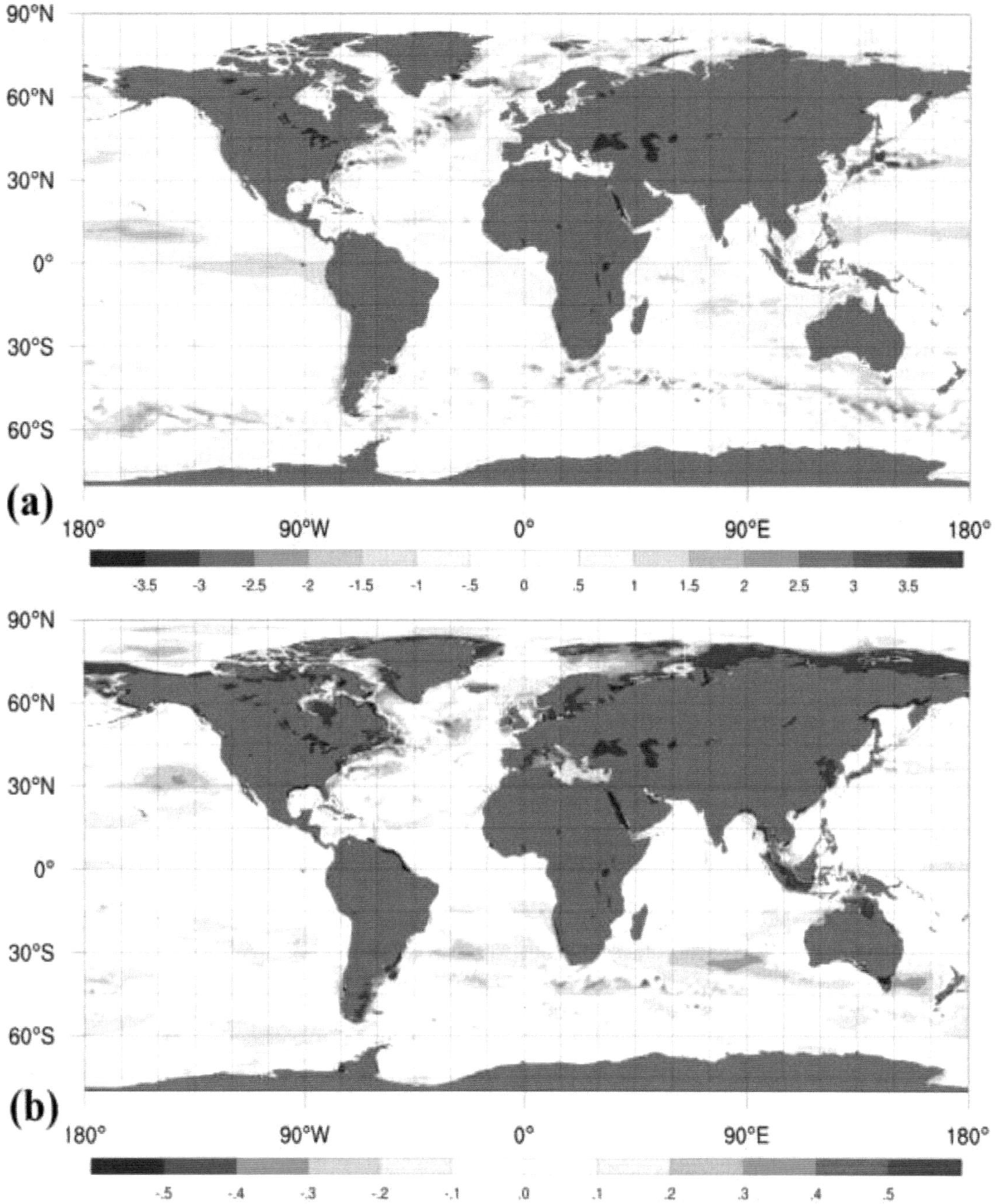

Figure 1. Temperature (a, °C) and salinity (b, psu) annual mean biases relative to the WOA09 data averaged over the upper 700 m ocean layer.

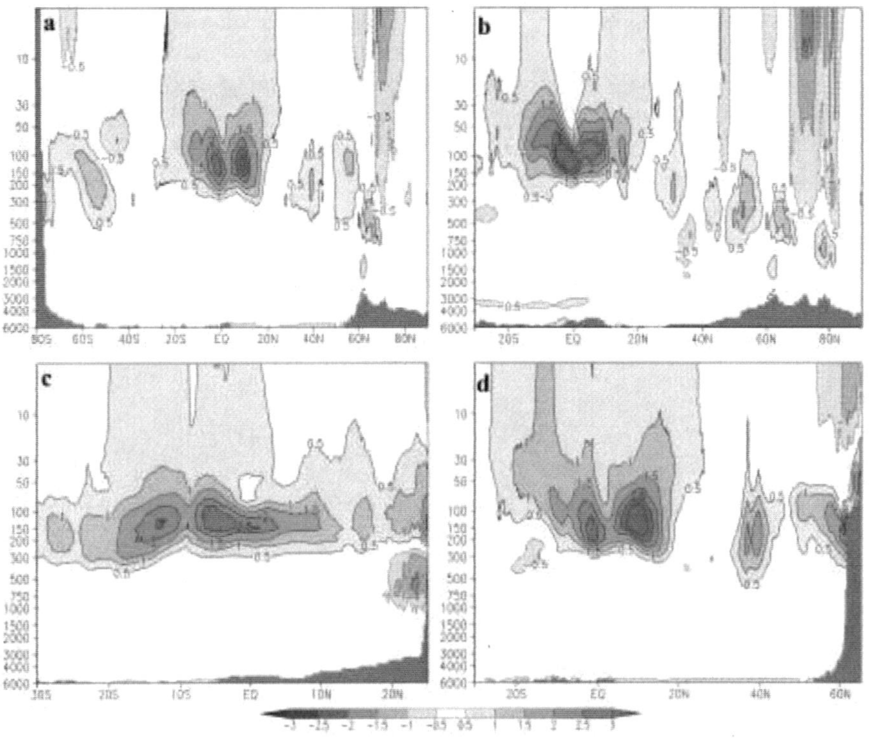

Figure 2. Annual mean temperature bias (°C) relative to the WOA09 data zonally averaged for the World (a), Atlantic (b), Indian (c), and Pacific (d) Oceans.

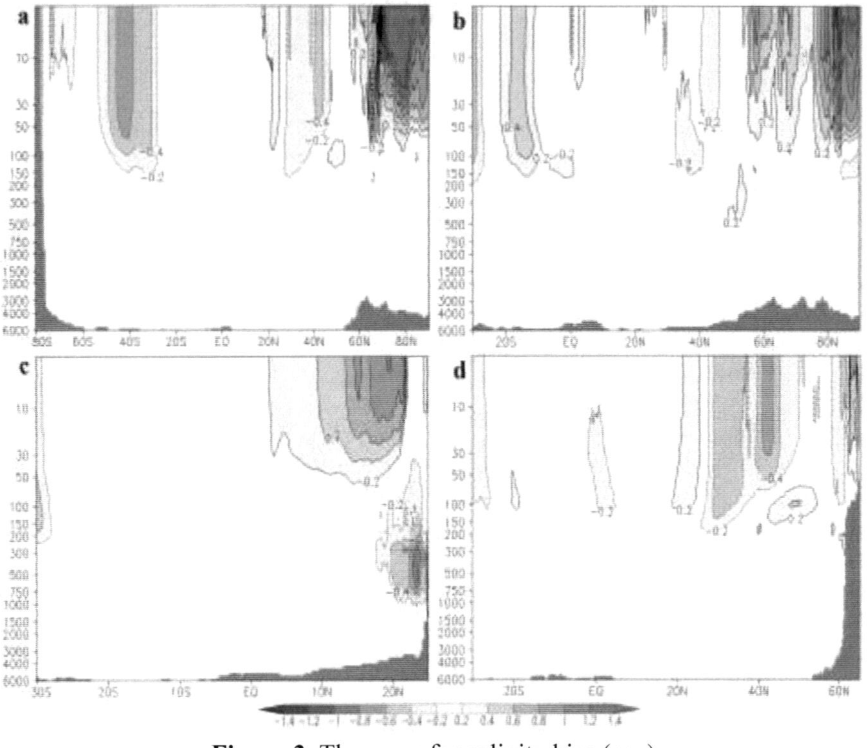

Figure 3. The same for salinity bias (psu).

To assess the quality of simulation for the annual cycle of ocean near-surface thermal characteristics, we consider the monthly mean surface temperatures and heat content (vertically integrated temperature) of the upper 250-metre ocean layer at a point of continuous ship observations, OWS Echo (35°N, 48°W). The monthly values of heat content are plotted against the monthly mean temperatures for the last year of the numerical experiment being considered, for the WOA09 monthly data, and also for the quasi-stationary solution of the INMIO model obtained in [4] with a resolution of 0.5°. The points of the hysteresis graph correspond to the months of the year, from January to December in the counterclockwise direction (Figure 4).

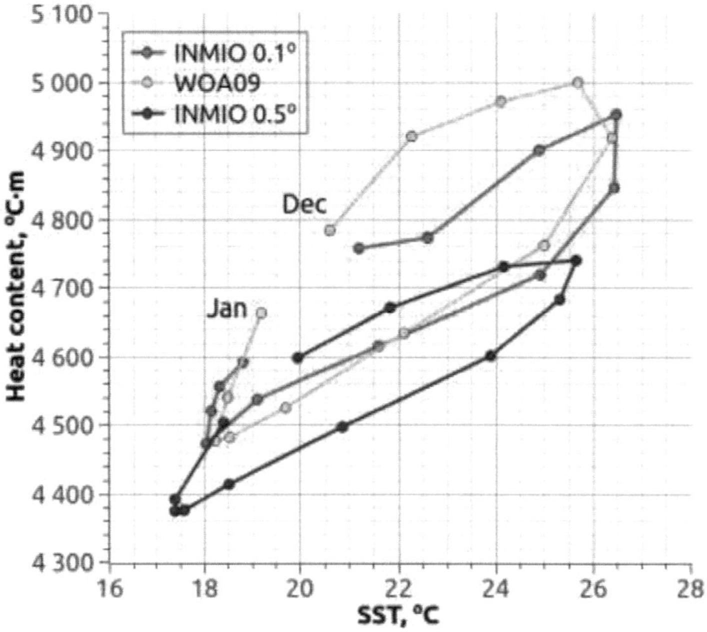

Figure 4. Seasonal heat content cycle at the OWS Echo site (35°N, 48°W) according to the WOA09 data and INMIO simulations of 0.5° and 0.1° resolution.

Both versions of the model reproduce well the phases of the seasonal heat content variability, while in [4] it was found that the phase errors in the model results often have the same order of magnitude as the differences between the climatological databases (in particular, between PHC3.0 and WOA2001). For the INMIO 0.5° configuration, significantly underestimated are the average heat content and its yearly amplitude, and, as for most of the models in [2], the maximum summer surface temperature. When using the model of 0.1° resolution, the first and third characteristics have almost coincided with the WOA09 climatology data. Since in the mid-latitudes the seasonal cycle of ocean heat content is primarily governed by the surface heat flux [22], better reproduction of the heat content is apparently associated with a better simulation of the upper ocean dynamics and, as a consequence, of the ocean-atmosphere interaction. In contrast to [2], it seems that correction of external forcing is not required. As for the autumn and winter heat content underestimated by the model, comparison with Fig. 1 allows us to conclude that it is likely caused by a shift of the warm Gulf Stream current to the north of the site being considered.

4. Conclusions

The numerical experiment performed within the framework of the CORE project is the first experience of analyzing the global thermohaline characteristics of the multi-year solution obtained using an eddy-resolving 0.1-degree INMIO model configuration. It has shown a noticeable improvement in the quality of the model solution in the upper ocean layer compared to the previous calculations with 0.5- and 0.25-degree configurations. However, the problem of the deviation of the Gulf Stream and the Kuroshio currents to the north after their separation from the continental slope, as well as the inaccurate representation of the North Atlantic current in the subpolar gyre region, has not been completely solved. A detailed analysis of the atmospheric forcing data and mixing parameterizations on the continental shelf will be required to improve the reproduction of the salt budget in the Arctic, and, possibly, in other marginal seas. Moreover, the analysis of deep ocean characteristics simulation will require much longer numerical experiments by further increasing the computational efficiency of the model program code.

Acknowledgements

This work was supported by the Russian Science Foundation (project no. 14-27-00126) and performed at the Institute of Numerical Mathematics, Russian Academy of Sciences.

References

[1] Brunet G, Jones S and Ruti P M 2015 *Seamless Prediction of the Earth System: From Minutes to Months* (Geneva: World Meteorological Organization)
[2] Griffies S M, Biastoch A, Böning C et al 2009 Coordinated ocean-ice reference experiments (COREs) *Ocean modelling* **26** (1-2) 1–46
[3] Danabasoglu G et al 2014 North Atlantic Simulations in Coordinated Ocean-ice Reference Experiments phase II (CORE-II). Part I: Mean States *Ocean Modelling* **73** 76–107
[4] Ushakov K V, Ibrayev R A and Kalmykov V V 2015 Simulation of the World Ocean climate with a massively parallel numerical model *Izv. Atmos. Ocean. Phys.* **51**(4) 362–80
[5] Ushakov K V, Ibrayev R A and Grankina T B 2016 Modeling the water circulation in the North Atlantic in the scope of the CORE-II experiment *Izv. Atmos. Ocean. Phys.* **52**(4) 365–75
[6] Ibrayev R A, Khabeev R N and Ushakov K V 2012 Eddy-resolving 1/10° model of the World Ocean *Izv. Atmos. Ocean. Phys.* **48**(1) 37–46
[7] Smith R D, Maltrud M E, Bryan F O and Hecht M W 2000 Numerical simulation of the North Atlantic Ocean at 1/10° *J. Phys. Oceanogr.* **30** 1532–61
[8] Chassignet E P and Marshall D P 2008 Gulf Stream Separation in Numerical Ocean Models *Ocean Modeling in an Eddying Regime* eds M W Hecht and H Hasumi (Washington, D. C.: American Geophysical Union)
[9] Chang Y S, Garraffo Z D, Peters H and Özgökmen T M 2009 Pathways of Nordic Overflows from climate model scale and eddy resolving simulations *Ocean modeling* **29** 66-84
[10] Kalmykov V V and Ibrayev R A 2013 A framework for the ocean-ice-atmosphere-land coupled modeling on massively-parallel architectures *Vychislitel'nye Metody i Programmirovanie* **14** 88-95
[11] Kalmykov V V 2013 Program complex for numerical modelling of the coupled system of ocean-atmosphere on massively parallel computers, PhD thesis (Moscow: INM RAS, in Russian)
[12] Locarnini R A *et al* 2010 World Ocean Atlas 2009, Volume 1: Temperature ed S Levitus *NOAA Atlas*
[13] Antonov J I *et al* 2010 World Ocean Atlas 2009, Volume 2: Salinity Ed S Levitus *NOAA Atlas*
[14] Large W and Yeager S 2004 Diurnal to decadal global forcing for ocean and sea-ice models: the data sets and flux climatologies *NCAR Technical Note: NCAR/TN-460+STR* (CGD Division of the National Center for Atmospheric Research)
[15] Schrum C and Backhaus J 1999 Sensitivity of atmosphere-ocean heat exchange and heat content in North Sea and Baltic Sea. A comparative Assessment *Tellus* **51**A 526–49
[16] Zalezak S T 1979 Fully multidimensional flux-corrected transport algorithm for fluids *J. Com. Phys.* **31** 335–62
[17] Killworth P D, Stainforth D, Webb D J and Paterson S 1991 The development of a free surface

Bryan-Cox-Semtner model *J. Phys. Oceanogr.* **21** 1333–48
[18] Kalmykov V V and Ibrayev R A 2013 Algorithm with overlaps for solving shallow water equations system on distributed memory parallel computers *Vestnik UGATU* **5**(58) 252–59
[19] Murray R J 1996 Explicit Generation of Orthogonal Grids for Ocean Models *J. Comp. Phys.* **126**(2) 251–73
[20] ETOPO5 1988 Data Announcement 88-MGG-02. Digital relief of the Surface of the Earth (Colorado: NOAA, National Geophysical Data Center)
[21] Ushakov K V, Grankina T B, Ibrayev R A and Gromov I V 2016 Simulation of Arctic and North Atlantic ocean water and ice seasonal characteristics by the INMIO-CICE coupled model *IOP Conference Series: Earth and Environmental Science* **48** 012013
[22] Sarmiento J L 1986 On the North and Tropical Atlantic heat balance *J. Geophys. Res.* **91** (C10) 11677–90

Sun's effect on the surface temperature in the Northern Hemisphere

V A Tartakovsky[1], N N Cheredko[1], Y V Volkov[1,2] and V G Maximov[1]

[1]Institute of Monitoring of Climatic and Ecological Systems SB RAS Akademichesky av. 10/3, Tomsk, 634055, Russia
[2]Tomsk Polytechnic University, Lenin av. 30, 634050, Tomsk, Russia
sun-effect@list.ru

Abstract. Series of temperatures measured at 818 weather stations in the Northern Hemisphere and one series of Wolf numbers for the period from 1955 to 2010 are considered. The sun's effect on the surface temperature is estimated by the cosine of the angle between two vectors determined by the series under study for the same months, year by year in the observation interval. The sun's effect is distributed over the weather stations, both in space and time. It has a quasi-monotonic dependence on the average monthly temperature, and reflects the geography of the climate, including the effect of warm ocean currents. Various manifestations of the sun's effect on the surface temperature have been discovered, and the conditions of their occurrence are discussed. The approach proposed in this paper can be used in observational data analysis, analytical transformations, and climate simulations.

1. Introduction

The existence of the solar system suggests that many natural and climatic processes on the Earth are controlled by the Sun. The simulation and forecasting of these processes are limited by the insufficient available body of knowledge, although scientists have studied the solar effect on climate elements for several centuries. The problem of solar-terrestrial relations is complex. We will consider only a few of the numerous aspects.

Along with its direct impact, the Sun regulates other effects on the geospheres. It modulates the flows of cosmic charged particles and the solar wind, which then change the conditions of energy conversion in the troposphere [1-3]. As the Earth passes through the regions of the interplanetary magnetic field, it changes the radiation balance, which leads to changes in the temperature and pressure in the surface atmosphere, with large variability mainly in the ultraviolet part of the solar radiation spectrum which, due to photochemical reactions, regulates the formation of ozone in the stratosphere affecting the surface temperature [4-9].

There is a theory that the effect of solar forcing on the geosphere is insufficient and, therefore, it is not able to change the climate. However, there are mechanisms [1, 10-14] through which a small energy of the regulator can initiate substantial climatic processes.

The solid foundations for climate simulation are fully developed in forms of systems theory, mathematical physics, computer sciences, etc. [15]. Mathematical models that allow studying various climate scenarios have been constructed, and significant results have been obtained. Nevertheless, this is not enough for practical forecasts. Some well-known challenges are: numerical instability of the equations in combination with inadequate initial or boundary conditions and coefficients, uncertainties in the external influences on the climate system, and the presence of many other unknown mechanisms [16]. There is also a need for assumptions about the stochastic properties of the processes forming the time series being analysed [17]. The mechanisms of solar energy transformations in the atmosphere are still subjects of debate. The physical nature of the solar-terrestrial connections has not been fully understood yet, and additional research is required for its clarification.

Content from this work may be used under the terms of the Creative Commons Attribution 3.0 licence. Any further distribution of this work must maintain attribution to the author(s) and the title of the work, journal citation and DOI.
Published under licence by IOP Publishing Ltd

The natural and climatic processes are manifestations of a complex combination of various factors, and it is almost impossible to identify and estimate all of them. Because of this complexity, the natural processes are fundamentally different from the processes in technical systems which are specially designed to implement some beneficial factors. For these reasons, representing the climate system by means of traditional physical quantities is not a simple task and can be somewhat ineffective.

At this stage, it is advisable to expand the phenomenological foundations of studying and forecasting climate, as well as to allocate the predictive patterns from the properties of observed signals by using general physical principles. In this connection, our goal is to identify such components of the series of observations that can accentuate the patterns of the climate system's response to external forcing.

Tartakovsky's papers [18, 19] based on an analysis of natural temperature series show that synchronicity is an essential property of climatic processes. The so-called initial product moments are used as a measure of synchronicity. Some opposite components of temperature and solar activity with extreme synchronicity have been found and investigated. A phase of temperature oscillations as a measure of their synchronicity was introduced. The cross phasing of temperature series was geographically localized, which leads to an adequate climate classification in the Northern Hemisphere [20].

This paper presents new results on the solar influence on the temperature in the surface layer of the atmosphere taking into account the location of weather stations, under the assumption that solar activity and associated cosmic influences can be characterized by a unique series of Wolf numbers. The Wolf numbers are determined by the calculation of sunspots, which astronomers have been performing for the last 200 years.

2. Data series and calculations

When performing the calculations, we used consecutive average monthly temperatures measured from 1955 to 2010 at 818 weather stations in the Northern Hemisphere. The data were obtained from the University of East Anglia website [21]. A series of average monthly Wolf numbers during the same period was obtained from the website of the Pulkovo Observatory [22]. For each of the 818 weather stations, the temperature series contained N=56×12=672 counts and the sample size was 672×818=549696. The Wolf number series also contained 672 counts.

The considered series of temperatures $x_{l,k}$ in degrees Celsius and the series of Wolf numbers s_k (dimensionless) represent the processes under study in an interdependent one-to-one manner. These series are real and have a finite energy. Here k is a discrete argument characterizing the time required in order to take N values in the observation interval, and l is the number of the series and the weather station where it was obtained.

By construction, the series under consideration are a sequence of values of smoothed natural processes at discrete moments of continuous time. In the solar system, quasi-periodic motions occur, and this is the reason for the analysis of climatic data in the same months, year by year in the observation interval. In this case, the values of the series of average monthly Wolf numbers and temperatures are selected and used to form a new series by calculating the index k as follows:

$$k = m + 12(j - 1);\quad m\,\mathrm{O}[1,12];\quad j\,\mathrm{O}[1,56]; \tag{1}$$

where m is the month's number and j is the year's number. Then the matrix of the average multi-year temperature at the weather stations with the number $l\,\mathrm{O}[1,818]$ for each month m of the 56 years was calculated according to the traditional formula:

$$T_{l,m} = \frac{1}{56}\sum_j x_{l,\,m+12(j-1)}\,. \tag{2}$$

The series under consideration can be continued over the entire real axis in a periodic even fashion. For such series, using the discrete Fourier transform, the real periodic spectra $X_{l,\nu}$ and S_ν were calculated. To equalize the lengths of the series and compare the spectra, the counts were preliminarily interpolated to $N = 2^{12}$ in both cases.

Let $\nu \in [1, N]$ be the discrete frequency. Then we can estimate the spectral density P_ν for all 818 temperature series, as well as the square of the modulus of the Fourier spectrum W_ν, which replaces the spectral density for the series of Wolf numbers:

$$P_\nu = \sum_{l=1}^{818} X_{l,\nu} \cdot X_{l,\nu}^*; \; X_{l,\nu} = \frac{1}{N}\sum_{k=1}^{N} x_{l,k} \exp(-i2\pi\nu k/N); \; P_\nu = P_\nu / \max(P_\nu);$$
$$W_\nu = S_\nu \cdot S_\nu^*; \; S_\nu = \frac{1}{N}\sum_{k=1}^{N} s_k \exp(-i2\pi\nu k/N); \; W_\nu = W_\nu / \max(W_\nu), \qquad (3)$$

where i is the imaginary unit and the asterisk denotes complex conjugation, although in this case it is redundant, since the spectra are real functions.

The series introduced in accordance with Eq. (1) can be considered as 12 temperature vectors $\boldsymbol{x}_{l,m}$ for each weather station with number l O[1, 818] and 12 vectors of Wolf numbers s_m for all weather stations. Initially, each vector had 56 components in accordance with the duration of the observation interval:

$$\boldsymbol{x}_{l,m} = \{x_{l,m+12(j-1)}\}; \; s_m = \{s_{m+12(j-1)}\}; \; j = 1, 2, ..., 56. \qquad (4)$$

For these vectors, we calculated the normalized initial product moments for each month and each weather station. This moment is the cosine of the angle between the vectors $\boldsymbol{x}_{l,m}$ and s_m. This value denoted by $\text{æ}_{l,m}$ characterizes the external cosmic and solar effect on the earth's surface temperature. The calculation of $\text{æ}_{l,m}$ can be performed in the following way:

$$\text{æ}_{l,m} = \cos(\boldsymbol{x}_{l,m}; s_m) = \sum_j x_{l,k} s_k \Bigg/ \left(\sum_j x_{l,k}^2 \cdot \sum_j s_k^2 \right)^{1/2}, \qquad (5)$$

where the index k is calculated as in Eq. (1). Evidently, the smaller the acute angle between the vectors $\boldsymbol{x}_{l,m}$ and s_m, the greater the positive sun's effect on the surface temperature; and the larger the obtuse angle, the greater the negative sun's effect.

When calculating the sun's effect as in Eq. (5), there is no centring, and there may be a non-zero mean value in the series, which may have a physical meaning. An example is the constant component of solar forcing. It is interesting that the normalization in Eq. (5) eliminates a coefficient independent of the time index (k or j) in the series. This coefficient could be used to describe the latitudinal dependence of solar forcing. However, it is not obvious that the Wolf numbers are related to geometry; they must be considered as an integral indicator of external forcing acting on the Earth as a whole. In this context, they characterize the only external input to the climate system, and the reactions of all of its internal mechanisms are displayed at the weather stations. No assumptions are made about any properties of the system. Only the interrelationships between the inputs and outputs are investigated using the completely correct tool, see Eq. (5).

3. Results and discussion

In Figure 1, the spectral densities W_ν and P_ν are compared for the cases of time according to Eq. (1) and of discrete continuous time. In the first case, the spectra W_ν and P_ν rapidly decrease, and it is clear that they have a common frequency band both in January and June. In the second case, the spectra W_ν and P_ν slowly decay, and, thus, aliasing above the Nyquist frequency is possible. They also have a weak intersection of the information-significant frequency intervals. For these reasons, the search for interrelations is more promising in the case of time calculated by Eq. (1), since this approach eliminates the periodicity and consolidates the information about the deviations occurring at present and those that have occurred in the course of many previous cycles. In the case of discrete continuous time, it is expedient to study a spectral interval in the vicinity of the carrier frequency associated with the annual cycle. Both variants were developed in [18-20].

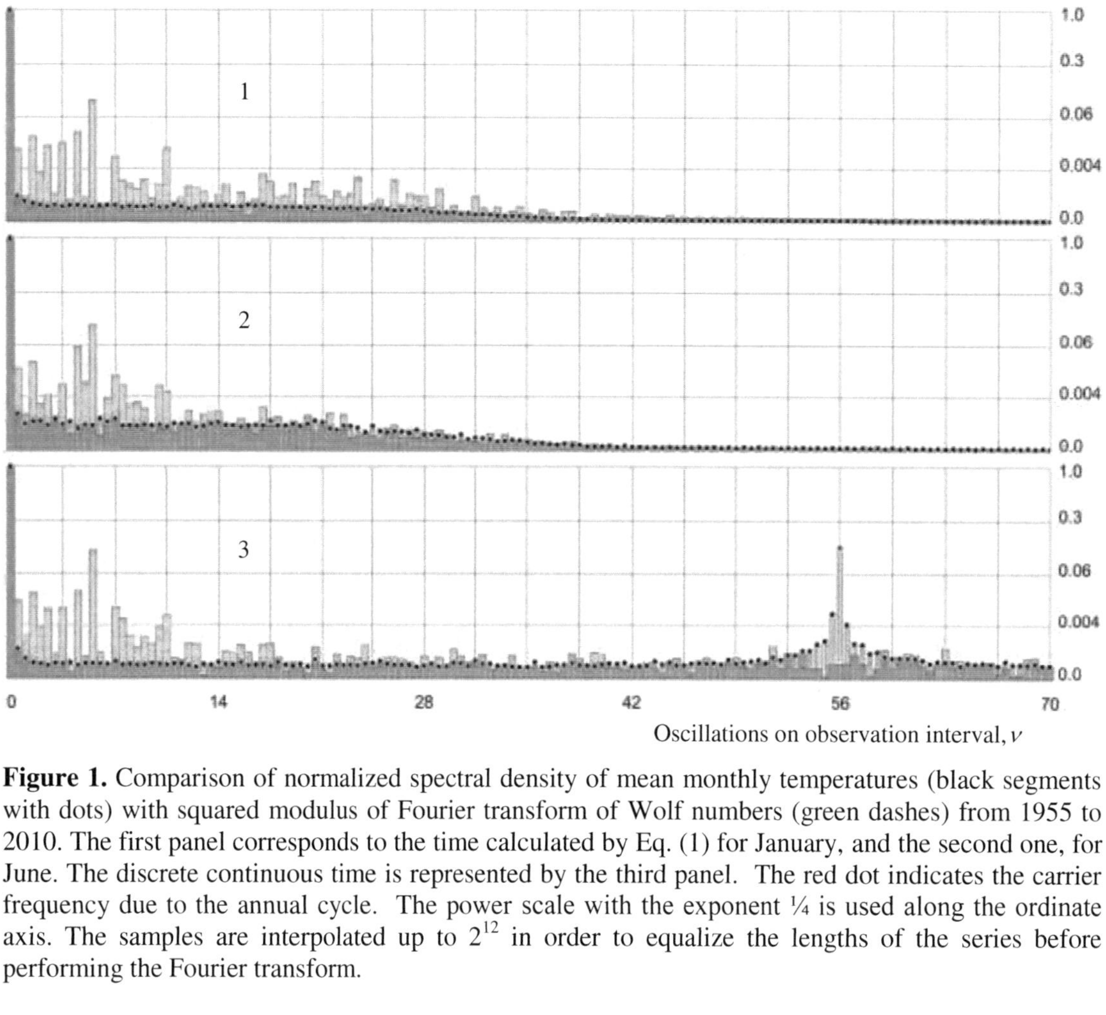

Figure 1. Comparison of normalized spectral density of mean monthly temperatures (black segments with dots) with squared modulus of Fourier transform of Wolf numbers (green dashes) from 1955 to 2010. The first panel corresponds to the time calculated by Eq. (1) for January, and the second one, for June. The discrete continuous time is represented by the third panel. The red dot indicates the carrier frequency due to the annual cycle. The power scale with the exponent ¼ is used along the ordinate axis. The samples are interpolated up to 2^{12} in order to equalize the lengths of the series before performing the Fourier transform.

3.1. The sun's effect on the surface temperatures for each weather station in each month

We will analyse the relationship between $æ_{l,m}$ (see Eqs. 1 and 5) and the average multiyear temperature $T_{l,m}$, which are functions of the index parameters: l is the number of the weather station and m is the month. A point-portrait of the compared values turns out to be very informative and is shown in Figure 2.

The spread in the points on the graph is quite small, such that a monotonic stepwise approximation is adequate. Some of its asymmetry is a consequence of the different numbers of points at negative and positive temperatures. There are fewer points to the left of zero and more to the right. It is important to note that the argument of this simple function in Figure 2 is the temperature $T_{l,m}$, which is quite complex distributed both spatially (index l) and temporally (index m).

Near zero, there is certain linearity in the arrangement of points, especially in the interval of ±1.5°C. Within this range, a linear regression was performed (nrms = 0.6% for the temperature axis). The slope of the straight line with respect to the abscissa is 7.5. This roughly corresponds to 1.4 rad. Therefore, within the regression interval of ±1.5°C, the temperature rise by one degree is caused by an increase in the value of the sun effect $æ_{l,m}$ by approximately 1.4 times and it reaches about ±0.3 at the boundaries of the interval.

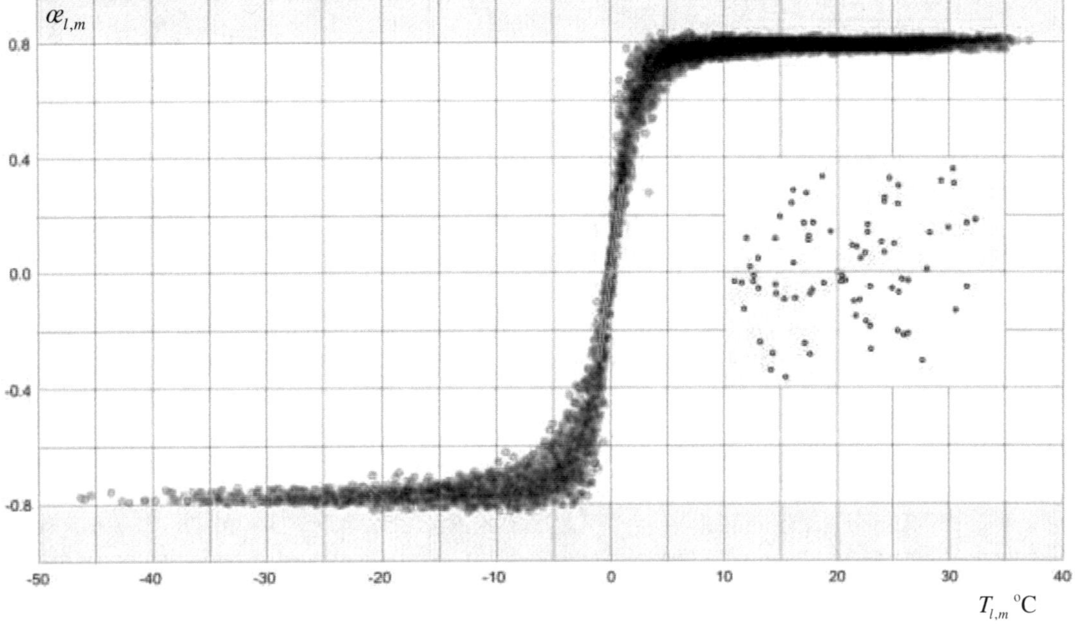

Figure 2. Sun's effect on the surface temperature as a function of the average multi-year temperature $T_{l,m}$ °C within the interval from 1955 to 2010. In total, there are 9816 points on the graph, of which 1977 are at a negative temperature and 7839 are at a positive temperature. The regression line is drawn in the range of ± 1.5°C, within which 535 points are located. The solid grey colour fills the range beyond ±0.8. Enlarged neighbourhood of the origin is in the inset panel. All 35 points in the even quadrants are located in the temperature range from about –0.077 to 0.083°C, and they correspond to the sun's effect from –0.201 to 0.289.

We note that the series of temperature and Wolf numbers are not centered, there is no such operation in Eq. (5); therefore, the location of points in the neighborhood of zero in Figure 2 can only be explained by some physical interaction of the processes under study. In particular, a linear change in the solar effect as a function of temperature in the interval about ±1.5°C suggests that a manifestation of phase transitions of water in the surface atmosphere is displayed.

Let us look at the facts concerning this assumption, although, of course, more research is needed. The growth of the sun's effect and the forced transition of water from the solid state to the liquid state do not lead to an increase in its temperature, since all energy is expended on breaking the intermolecular bonds. When water freezes, a heat energy is released without lowering the temperature, and the sun's effect has a negative growth in Figure 2. These processes are symmetrical: the more energy is absorbed, the more is allocated. In general, the location of the points in Figure 2 does not contradict this. However, the temperatures of phase transitions depend on many parameters of water, on the moisture content of the surface air at different observation points, and this may affect the spread of points in the range of realization of the sun's effect on temperatures.

As follows from Figure 2, most of the 9,816 points are in the odd quadrants. This is why the signs of the sun's effect $æ_{l,m}$ and of the average monthly temperature $T_{l,m}$ coincide. Only 35 points are in the even quadrants. They are located near the origin of coordinates in a computed temperature range from about –0.077 to 0.083°C and for the sun's effect from –0.201 to 0.289. It is not yet clear what the reason is for this.

As follows from the Parseval theorem, the location of most of the points in the odd quadrants at extreme levels is equivalent to the coincidence of the signs of a large number of similar spectral harmonics with significant energy of the investigated processes. This is a manifestation of the synchronizing influence of the Sun on natural and climatic processes [19].

Beyond the interval of ±1.5°C and up to ±15°C, the values of $æ_{l,m}$ are gradually moving towards saturation of up to ±0.8. It occurs with increasing in both positive and negative temperatures. Therefore, beyond the interval of ±15°C, the maximum possible inherent consistency of the series of Wolf numbers with all the series of the average monthly temperature is realized.

The values of $æ_{l,m}$ are bounded as the cosine function in Eq. (5), but the absolute value of the saturation level is essentially less than one. This fact can be explained by the real ratio of values of the temperature and the Wolf numbers, also taking into account the physical meaning of these quantities. Here, we state only what is clear: the moduli of the sun's effect on the temperatures (5) in space and time are majorized by 0.8 already for the observed temperatures and not by a theoretically possible value equal to one.

3.2. Monthly variations of the sun's effect on the surface temperature

Let us perform a qualitative analysis of the sun's effect $æ_{l,m}$ on the surface temperature at the 818 meteorological stations in the annual course. As seen in Figure 3, the sun's effect can have positive values throughout the year and can change the sign from negative in a cold period to positive in a warm period. The range of changes, as already noted, is equal to approximately ±0.8.

The climatic conditions of the zones from the equatorial to the sub-tropical are known to be the most stable in the annual course. It turned out that at most of the weather stations of these zones (orange in Figure 3) the sun's effect has maximum values near a level of 0.8 throughout the year. However, in the period from November to March, at some stations these values decrease as compared to the top level. Such variability occurs at stations located in transitional zones with temperate climate.

For most weather stations in the temperate zone (black in Figure 3), the sun's effect values are clustered near a top level of 0.8 in the period from April to October. From November to March, the values of the sun's effect at these stations are lowered to a level of –0.8.

At the stations of the sub-arctic and arctic zones, the sun's effect is saturated to a bottom level of –0.8 during the winter months and during the transitional seasons, as well as during the polar night (black in Figure 3). In the summer months, the values of the sun's effect are close to a top level of 0.8.

There is a significant climate-regulating influence of large warm currents on the geosystem. The heat transferred by the water softens the climate of the regions so that they approach the conditions of the lower latitudes in terms of the temperature changes throughout the year. Based on the values of the sun's effect on the surface temperature, we selected those stations in the temperate zone which are most susceptible to the influence of the global warm currents. The following locations of weather stations are included: the coast of North-West Canada and South Alaska where the sun's effect is due to the warm Alaska Current (blue in Figure 3) and the weather stations influenced by the Gulf Stream (green in Figure 3). A list of these two groups of stations is given in [19].

There are also weather stations in the sub-tropical zone with negative values of $æ_{l,m}$ and $T_{l,m}$ in the winter months. These stations are characterized by high altitudes above sea level. The climatic conditions are more severe than in the surrounding territories. The type of relief and features of orography are well known to significantly affect the structure of temperature fields. These unusual stations are listed in Table A.1; the sun's effect changes are shown in black in Figure 3.

The delivery of solar energy to specific places occurs mainly through direct heating of the surface and atmosphere in accordance with the climatic zone, as well as heat transfer in accordance with the global and regional circulation processes. It has been stated previously [19] that the annual course of the second initial moments of the series of Wolf numbers' extreme components shows the existing geography of the climate. Note that the annual course of the sun's effect on the surface temperature in Figure 2 also displays this geography.

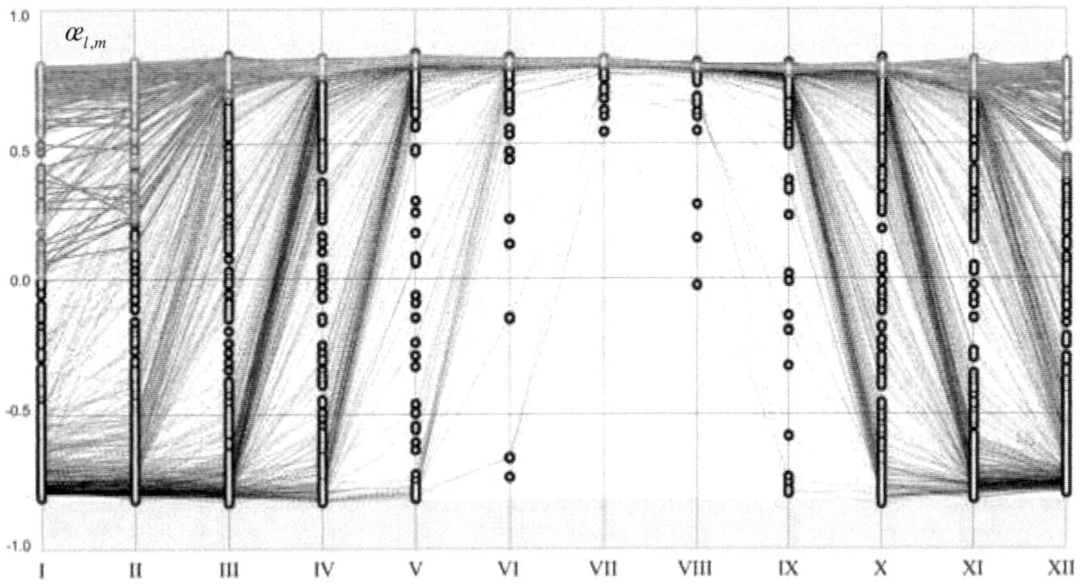

Figure 3. Monthly variations of the sun's effect on the surface temperature from 1955 to 2010 at weather stations with indices l. The yellow and orange colours represent 351 stations from –0.5°S to 45.8 N (degrees of latitude). The green colour marks 39 stations in the North Atlantic from 46 to 62.9 N. The blue and light-blue colours mark nine stations on the north-west coast of America from 45.6 to 55 N. The black colour marks 419 stations that do not coincide with the previous from 29.7 to 80.6 N. The months (the index m) are denoted by Roman numerals along the abscissa. The solid grey colour fills the range beyond ±0.8.

4. Conclusions

The sun's effect on the surface temperature was detected by applying a new estimation to the problem under study which corresponds to the cyclic motions in the solar system. In addition, a useful graphic display was used.

A series of Wolf numbers as an integral indicator of the external forcing and the only input signal to the earth's climate system was compared with many output signals, namely, temperature series measured at 818 weather stations of the Northern Hemisphere between 1955 and 2010. The investigation of the sun's effect on the surface temperature revealed new details of the global energy redistribution in the climate system.

◦ The sun's effect is distributed over the weather stations in space and time, and its dependence on the average monthly temperature can be approximated by a monotone function.

◦ In the range of ±1.5°C, a one-degree rise in temperature is caused by an increase in the sun's effect of approximately 1.4 times.

◦ The sun's effect is saturating on a constant level of about ±0.8, with an accompanying increase in the absolute value of temperature.

◦ The maximal sun's effect is approximately 80% of that which is theoretically possible, and this is achieved in ranges from approximately 2.5 to 45°C both for negative and positive temperatures.

◦ The sun's effect varies during the year in accordance with the geography of the climate and, in particular, the warm global ocean currents.

Acknowledgements
This research was supported by the Russian Academy of Sciences.

References
[1] Pudovkin M I 1996 Effect of solar activity on the lower atmosphere and weather *Soros Educ. J.* **10** 106–13
[2] Svensmark H and Friis-Christensen E 1997 Variation of cosmic ray flux and global cloud coverage – a missing link in solar–climate relationships *J. Atmos. Sol.-Terr. Phys.* **59** (11) 1225–32
[3] Courtillot V, Gallet Y, Le Mouël J-L, Fluteau F and Genevey A 2007 Are there connections between the Earth's magnetic field and climate? *Earth and Planet. Sci. Lett.* **253** (3) 328–39
[4] Karakhanyan A A and Molodykh S I 2015 Pressure field structure as the Earth passes the interplanetary magnetic field sectors *Sovr. Probl. DZZ Kosm.* **12** (4) 99–105
[5] Lean J, Rottman G, Harder J and Kopp G 2005 SORCE contributions to new understanding of global change and solar variability *Sol. Phys.* **230** 27–53
[6] Jager C and Usoskin I 2006 On possible drivers of Sun-induced climate changes *J. Atmos. Sol.-Terr. Phys.* **68** (18) 2053–60
[7] Rakipova L R 1974 *Vlijanie variacij koncentracii ozona na termicheskij rezhim atmosfery. Solnechno-atmosfernye svjazi v teorii klimata i prognozah pogody* (Leningrad: Gidrometeoizdat Press) 361 p.
[8] Callis L B and Nealy J E 1978 Solar UV variability and its effect on stratospheric thermal structure and trace constituents *Geophys. Res. Let.* **5** (4) 249–52
[9] Haigh J D, Blackburn M and Day R 2005 The response of tropospheric circulation to perturbations in lower stratospheric temperature *J. Clim.* **18** 3672–91
[10] Pudovkin M I 2004 Influence of solar activity on the lower atmosphere state *Intern. J. Geomagn. Aeron.* **5** (2) GI2007
[11] Scafetta N 2013 Solar and planetary oscillation control on climate change: Hind-cast. forecast and a comparison with the CMIP5 GCMS *Energy and Env.* **24** (3-4) 455–96
[12] Scafetta N 2014 Global temperatures and sunspot numbers. Are they related? Yes but non linearly. A reply to Gil-Alana et al *Physica A* **413** 329–42
[13] Lam M M, Tinsley B A 2016 Solar wind-atmospheric electricity-cloud microphysics connections to weather and climate *J. Atmos. Sol.-Terr. Phys.* **149** 277–90
[14] Svensmark J, Enghoff M B, Shaviv N J and Svensmark H 2016 The response of clouds and aerosols to cosmic ray decreases *J. Geophys. Res.: Space Phys.* **121** (9) 8152–81
[15] Dymnikov V P 2007 *Ustojchivost' i predskazuemost' krupnomasshtabnyh atmosfernyh processov* (Moscow: IVM RAN Press) 283 p.
[16] http://meteoinfo.ru/forecasts-limits (01.06.2017)
[17] Ghil M, Allen M R, Dettinger M D, Ide K, Kondrashov D, Mann M E, Robertson A W, Saunders A, Tian Y, Varadi F and Yiou P 2002 Advanced spectral methods for climatic time series. *Rev. of Geophys.* **40** (1) 1–41
[18] Tartakovsky V A 2015 Synchronous analysis of the Wolf numbers and temperature series from weather station in the Northern Hemisphere of the Earth *Atmos. Ocean. Opt.* **28** (2) 182–8
[19] Tartakovsky V A 2015a Synchronicity as an essential property of solar–terrestrial relations: latent components *Nonlin. Proc. in Geophys. Discus.* **2** 1275–99. DOI:10.5194/npgd-2-1275-2015
[20] Cheredko N N, Tartakovsky V A, Krutikov V A and Volkov Yu V 2017 Climate Classification in the Northern Hemisphere Using Phases of Temperature Signals *Atmos. Ocean. Opt.* **30** (1) 63–9. DOI: 10.1134/S1024856017010043

[21] http://www.metoffice.gov.uk; https://crudata.uea.ac.uk/cru/data/temperature) (01.06.2017)
[22] http://www.gao.spb.ru (01.06.2017)

Appendix A

Table A.1. Weather stations in the sub-tropical zone with negative values of the sun's effect on the surface temperature.

#	ISI	Latitude, °	Longitude, °	Altitude, m	Alpha-2 ISO 3166-1	*T*,°C, Jan
1	55591	29.7	91.1	3650	CN	−1.4
2	56137	31.2	97.2	3316	CN	−2.0
3	52836	36.3	98.1	3191	CN	−10.0
4	72486	39.3	−114.9	1909	US	−3.8
5	72564	41.1	−104.8	1872	US	−2.6
6	17096	39.9	41.3	1758	TR	−9.0
7	72576	42.8	−108.7	1696	US	−6.4
8	17170	38.5	43.4	1661	TR	−3.6
9	72569	42.9	−106.5	1612	US	−4.7
10	52533	39.8	98.5	1477	CN	−9.4
11	72476	39.1	-108.6	1473	US	−3.0
12	51828	37.1	79.9	1375	CN	−4.6
13	72578	42.9	−112.6	1358	US	−4.5
14	72583	40.9	−117.8	1315	US	−1.2
15	51709	39.5	76.0	1291	CN	−5.6
16	72572	40.8	−112.0	1289	US	−1.7
17	72666	44.8	−107.0	1209	US	−5.6
18	17092	39.7	39.5	1154	TR	−2.9
19	72465	39.4	−101.7	1124	US	−2.0
20	53614	38.5	106.2	1111	CN	−8.1
21	72677	45.8	−108.5	1091	US	−4.1
22	17244	37.9	32.5	1031	TR	−0.0
23	72662	44.1	−103.1	1030	US	−4.8

Mesoscale modelling of the summer climate response of Moscow metropolitan area to urban expansion

M I Varentsov[1, 2], P I Konstantinov[1] and T E Samsonov[1]

[1] Lomonosov Moscow State University, Department of Geography, Russia, Moscow, GSP-1, Leninskiye gory, 19991

[2] A.M. Obukhov Institute of Atmospheric Physic of Russian Academy of Science, Russis, Moscow, Pyzhevsky pereulok, 3, 119017

E-mail: mvar91@gmail.com

Abstract. In this paper, the experience of applying a regional climate model to simulating the summer climate features of Moscow metropolitan area is examined. Also, an assessment is made of climate response to the implementation of a scenario of twofold city expansion. The model (COSMO-CLM) was adapted to the conditions of the region under investigation, supplemented by specific urban canopy parameterization and equipped with realistic parameters of urban surface. It was possible to successfully simulate the summer meteorological regime of Moscow region and, specifically, the temporal and spatial variability of the Moscow urban heat island (UHI). First results of the simulation for the city expansion show that the new urbanized areas on the periphery of Moscow have a heating impact on its central part, which leads to an increase of the UHI effect, on the average, of 10% (in the central area). In extreme heat events the temperature response to this scenario is much stronger, which may deteriorate human health and increase thermal stress. The simulation results also show that the city of Moscow is characterized by a positive anomaly of summer precipitation (in the city and its leeward side it increases by 10-30%). The scenario of urban expansion enhances this anomaly by 5-10% and increases its area.

1. Introduction & motivation

Due to the current processes of climate change and urban growth, it is especially important to perform an investigation of local climate modifications caused by expansion of cities and changes in their planning and building features and of the contribution of such modifications to observed or predicted climate changes. It is well known that the features of urban surface strongly influence the atmospheric boundary layer and determine the microclimatic features of local environment, such as urban heat island (UHI) [1,2]. Also, UHI affects the atmospheric processes, such as the precipitation regime [3], mesoscale circulations [4], and the conditions of meteorological comfort of urban dwellers [5].

The Moscow megacity is the biggest monocentric agglomeration in Europe (population: 16-17 million people [6]). Due to its size, location in a zone of continental climate and specific building features, such as the predominance of high-rise block-houses, Moscow forms an intensive UHI with a mean annual intensity value equal to 2 °C (according to the Balchug weather station at the city centre) and with maximum values up to 12-13 °C [7]. Such values of UHI intensity are higher than those for a number of other big megacities including London [8,9], Beijing [10,11], and Shanghai [12]. All these features, in addition to the compact and relatively symmetric shape of the city and its location within a flat and homogeneous terrain makes Moscow a promising and interesting place for urban climate studies [13].

Another reason making such research especially important for Moscow is related to the trend of UHI intensification during the last decades. Since the 1970s, the mean **UHI intensity (UHII)**, defined

Content from this work may be used under the terms of the Creative Commons Attribution 3.0 licence. Any further distribution of this work must maintain attribution to the author(s) and the title of the work, journal citation and DOI.
Published under licence by IOP Publishing Ltd

as the difference between the temperature in the city and the **mean rural temperature** (calculated for 9 rural reference weather stations surrounding the city, see Appendix 1), has grown by 0.4 °C (from 1.6 °C to 2 °C), which could be caused both by the population growth (since that time the city population has approximately doubled) and climate change. In other words, urban-induced enhancement of global warming [14] takes place. For instance, warming for 1988-2016 in the city centre (Balchug weather station) is 0.55 °C/10 years instead of 0.36 °C/10 years for the mean rural annual temperature. For the mean summer (JJA) temperature the effect of urban amplification is even more intensive: 0.93 °C/10 years for the city centre instead of 0.55 °C/10 years for the rural area. This pattern seems to be especially dangerous because of increasing repeatability of high air temperatures, which has a negative impact on the citizens' well-being and health under the conditions of summer heat waves [5,15–17]. It is important that the highest warming rates are observed not at the weather stations located at the edges of the Moscow agglomeration, where active development of new territories has taken place during the last decades, but in the city center, where the land use changes are much smaller.

For these reasons, further investigation of the drivers of the observed climate changes and possible future changes becomes an especially important scientific problem. For the city of Moscow it becomes even more important because of ambitious projects of the further city development, such as plans of development of the so-called "New Moscow" area located to the south-west of the city, which was administratively joined to Moscow in 2011, and the program of renovation of old five-story buildings adopted in 2017 (according to it, more than 5000 buildings all over the city will be demolished and replaced by at least twice as high buildings).

However, more detailed investigation of the city growth and development effect on its climate based on observational data is very complicated because of the small number of stations with long-term datasets and the complicated problem of separating signals from different factors affecting urban-caused climatic anomalies, including the urban growth, local land-use modifications in the stations' surroundings, and the changes in the repeatability of different weather patterns under the climate change. Very promising for such investigations are mesoscale regional climate models (RCMs), which allow one to simulate changes of selected parameters under invariance of other conditions. This study is devoted to application of such model for understanding of climatic consequences of a possible two-fold expansion of the Moscow city for the summer season.

2. Data and methods

2.1. Configuration of numerical experiments with RCM

This study is based on numerical simulations with a regional climate model called COSMO-CLM (Climate Limited-area Modeling community). It is a nonhydrostatic model of the atmosphere and soil active layer based on the COSMO model (COnsortium for Small scale MOdelling) and adopted for long-term numerical experiments [18]. This model was applied for three-step dynamic downscaling of ERA-Interim reanalysis [19] data (with a long. and lat. grid step of 0.75°), which was used as the initial and boundary conditions for running the model for a "basic" domain D1. During the calculations the boundary conditions were undated 4 times per day to keep meteorological fields in the model close to reality. In addition, to ensure a more reliable binding of the internal model mode to the real atmospheric dynamics, the spectral nudging technique [20] was used. The importance of its use in such applications is shown in [21,22]. It should be noted that the dynamic downscaling method used in this study is the most complicated and computationally expensive one among all methods of downscaling used in meteorology and climatology. However, instead of others it provides not only statistical information for selected variables, but detailed 4-dimensional (spatio-temporal) fields of all meteorological variables (available in the model) calculated with consideration of all possible processes and their interactions, which could be reproduced within the framework of the model's physics and numerics.

Domain D1 covers the central part of European Russia with a size of approx. 1700x1700 km and a 12-km horizontal grid step (140x140 grid cells). Modelling results for D1 were used as the initial and boundary conditions for intermediate domain D2 with a size of 600x600 km (200x200 grid cells) and a 3-km grid step covering the Moscow region and the neighboring regions of Russia. Then the same procedure was repeated for domain D3 with a size of 180x180 km with a 1-km grid step. The final 1-km step was selected to explicitly resolve large features of the urban landscape and the urban-caused mesoscale circulations and avoid the problems of parameterization of atmospheric turbulence for smaller grid steps [23]. For D2 and D3 domains the spectral nudging was not used, the boundary conditions were updated more often (every hour), and a shallow convection scheme was used instead of a deep convection parametrization [24] used for D1. The time step was defined as 120 sec. for D1, 40 sec. for D2, and 10 sec. for D3. For all domains 50 vertical levels in the atmosphere and 11 levels in soil were used. Surface features such as albedo, leaf index, soil type, etc. were set according to Globcover database [25].

Model options were tuned to fit the modelled meteorological regime to observations based on a large number of preliminary test experiments (details are not shown). In general, the procedure of model tuning was aimed to solve the problem of significant overestimation of nocturnal temperatures and underestimation of its diurnal variations by the model, which is also reported in [26–28], as well as the problem of temperature overestimation for hot and dry summer seasons [29]. Details are listed in appendix 2.

The numerical experiments configured as described above were initialized on the 1st of May and were run until the 31st of August. The first month (May) was considered as a spin-up period required for adaptation of the model to the meteorological regime. Such experiments were performed for five summer seasons: 2009, 2013-2015, and the extremely hot summer of 2010 [29,30]. 2014 was also rather hot and dry, and the other three years were close to the climate means (Table 1).

2.2. Description of urban surface features for modern city and scenario of its extension

A bulk urban canopy model (UCM) called TERRA_URB [31,32] coupled to COSMO-CLM was used to simulate the urban climate features. To take into account the features of heat balance of the urban surface including the influence of "street canyons" [33], TERRA_URB provides semi-empirical corrections of surface parameters (roughness length, albedo, emissivity, etc.), parameterizes the moisture balance of impervious surface, and realizes the so-called tile approach, which assumes the existence of urbanized and natural parts within each model cell. The anthropogenic heat flux (AHF) was calculated according to a parameterization in [34] which takes into consideration its "typical" diurnal and annual variation and the dependence of its annual amplitude on the latitude.

Running of this UCM requires the definition of parameters describing the urban surface: urban area fraction F_{urb}, mean annual value of AHF, building height H, building area fraction within urban fraction F_{urb}^{bld} , and the aspect ratio of street canyons H/W, where W is the mean street width. Such parameters for modern city were obtained with the use of GIS-processing of OpenStreetMaps data according to [35]. Mean values of H and H/W within each model cell were explicitly calculated by GIS algorithm with a 1-km grid step. It should be noted that H values were calculated from the number of buildings' stories available in OpenStreetMaps, with an assumption that height of one story is 3 m and the height of basement and roof is 2 m. For buildings with unknown heights, typical values were used. The urban fraction F_{urb} was calculated as a combination of output parameters of OpenStreetMaps data GIS-processing:

$$F_{urb} = \min[1 - F_{nat}, \max(F_{bld} + F_{road} + k_{ind}F_{ind} + k_{res}F_{res}; F_{bld} + F_{can})],$$

where F_{nat} is the known fraction of natural surfaces (open water and vegetation), F_{bld} and F_{road} is the buildings and roads fraction, F_{ind} and F_{res} is the fraction of industrial and residential areas, and F_{can} is the fraction of street canyons identified by GIS. The ratio of artificial and impervious surfaces within the industrial and residential (usually the courtyard) areas was controlled by k_{ind} and k_{res} coefficients, which were set to 0.5. The annual mean value of AHF was defined according to [36],

IOP Conf. Series: Earth and Environmental Science **96** (2017) 012009 doi:10.1088/1755-1315/96/1/012009

which gives an estimate of its mean value for Moscow as 75 W/m2 *for urbanized area* in agreement with [37,38]. According to the approximation of its annual variation [34] used in the model the mean value for the summer season is about 30 W/m2. To take into account the dependence of AHF on the building height and density, the following formula was used for its calculation for a certain model grid cell:

$$AHF = \min(\overline{AHF} \cdot F_{urb} \cdot \frac{F_{urb}^{bld} \cdot L_{bld}}{M}, AHF_{max}),$$

where F_{urb}^{bld} is the building fraction within the urban fraction, L_{bld} is the mean number of building stories, $\overline{AHF} = 75\, Wt/m^2$ is the adopted estimate of annual mean AHF averaged over the urbanized area within Moscow, $M = \frac{\sum(F_{urb} \cdot F_{urb}^{bld} \cdot H_{bld})}{\sum F_{urb}}$, $AHF_{max} = 150 \frac{Wt}{m^2}$ is the adopted maximum value of AHF for a model cell.

The considered scenario of urban expansion was developed according to the assumptions of a twofold increase in the population of the Moscow agglomeration, preservation of the current pattern of its development, keeping unchanged the building and planning features of the "old" city part within the Moscow Automobile Ring Road (MKAD), and development of new areas outside this road. The urban surface parameters required for the model and representing this scenario were obtained by a developed iterative algorithm. During each step of this algorithm new built model cells are created near previously built cells with a prescribed probability, which models gradual growth of the city. The work of the algorithm continued until the sum of the product of the area of buildings by the number of stories doubled (it was assumed that the population is proportional to this parameter). The spatial distribution of the urban fraction and building height for the modern city and the considered scenario of its expansion is shown in Figure 1. It should be noted that the proposed scenario of urban expansion does not pretend to be realistic, but it is convenient for studying the effect of urban growth on local climate.

The TERRA_URB scheme was used only for the "final" domain D3, for which numerical experiments with a realistic modern city (URB) and a double city (URBx2) were performed for each summer season. In addition, an experiment without the TERRA_URB scheme (noURB) assuming the replacing of urban areas by natural landscapes was also calculated to compare the climate response to city expansion with climatic anomalies caused by the influence of modern city within the same modelling framework.

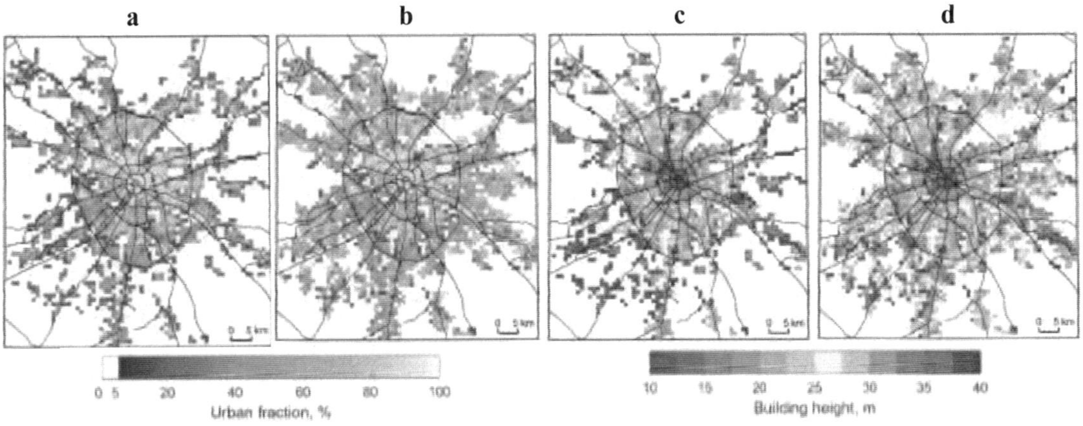

Figure 1. Spatial disrtibution of urban fraction (a, b) and building height (c, d) defined for model cells for URB (a, c) and URBx2 numerical experiments.

2.3. Observations used for model verification

The use of the model to study the phenomenon-induced response to some changes requires checking the ability of the model to simulate this phenomenon (i.e., verification of the model). For this purpose we used observations at meteorological stations located in the Moscow region, including new automatic meteorological stations (AWS) and automatic air-quality stations (AAQS) of MosEcoMonitoring (in total, more than 70 measurement points in the city and its 100-km neighborhood). A more detailed description of the meteorological observations used to verify the model can be found in [7, 39–41]. Despite the fact that the AAQSs do not fully correspond to the meteorological standards, a comparison of the observation data of such stations installed at the same points as the standard meteorological stations showed satisfactory results for the mean daily and mean nocturnal air temperatures [39]. The main WMO stations that characterize different parts of the Moscow city are Balchug station in the city center (600 meters from the Kremlin) and the Meteorological Observatory of Moscow State University (MSU campus) and the VVC weather station located in park zones (see Appendix 1 and Figure 3a).

3. **Results and discussion**

3.1. Results of model verification

Comparison of the observational data at meteorological stations, AWS, and AAQS with the modeling results show that COSMO-CLM simulations for conditions of the modern city adequately reproduce the meteorological regime of the Moscow region and the features of the UHI including the mean rural temperature and UHII in the city center and urban parks, which is confirmed by the model vs observations plots (Figures 2a-c).

The simulated spatial variability of the temperature within the city is also in good agreement with the observations, which is confirmed by the scatter plots for the mean values during the summer of 2014 to observation (at all available meteorological stations, AWS, and AAQS) and modeling data (Figures 2d-e). Details of the methodology for comparing observations and modeling data are described in [39]. In addition, the results from [39] show that the model adequately simulates the vertical extent and structure of the UHI for the other considered summer seasons.

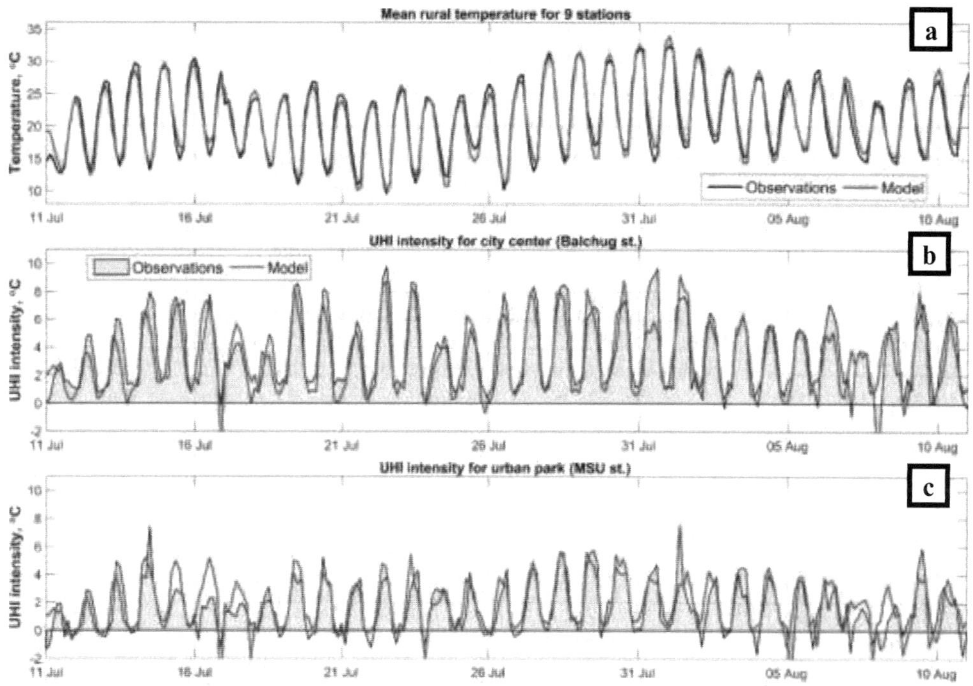

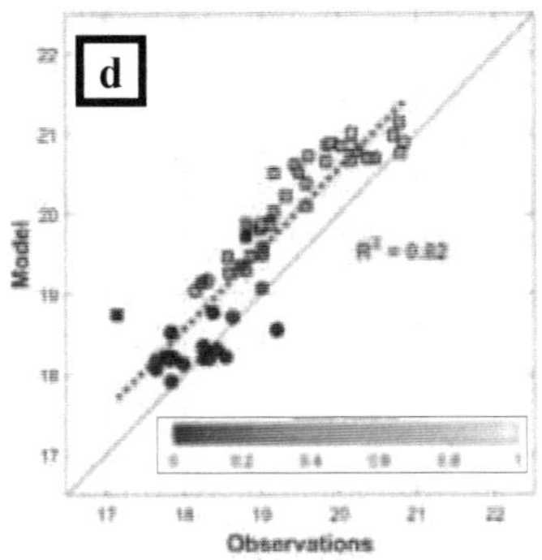

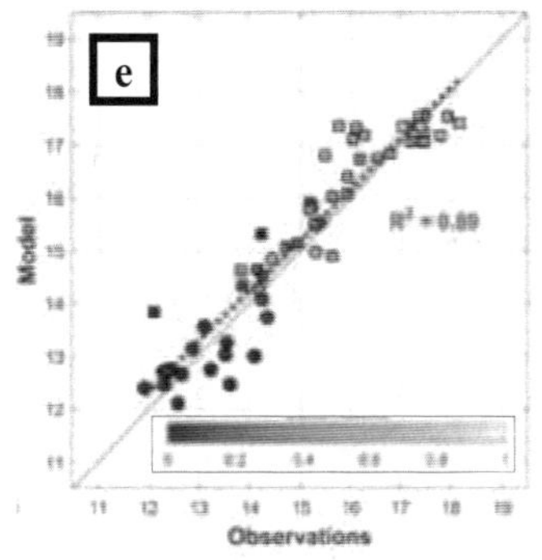

Figure 2. Temporal variations of modeled and observed values of mean rural temperature (a), UHII for city center (b) and urban park (MSU) (c) during 11 June – 11 August 2014; scatterplots for mean (d) and mean nocturnal (at 0 UTC) (e) temperatures for 2014 summer season at different points within 110 km from city center according to observations (weather stations, AWSs, and AAQSs) and modeling results. In (d, e) the color of the markers represents the urban fraction for corresponding model cell, square markers are used for points within 25 km from the center, and circles, outside this distance.

For the other summer seasons considered the model also adequately simulates the temperature regime, UHII and precipitation amount, which is confirmed by statistics of the model-to-observation comparison in Table 1 (mean error ME, root mean square error RMSE, and correlation coefficient R). It should be noted that for precipitation sums only the mean error, ME, was calculated, because more detailed verification of mesoscale precipitation fields based on observations at weather stations is a complicated problem. The ME was calculated as the difference between the modelled precipitation sum averaged over D3 domains (P_3 value from Table 2) and its mean values over the weather stations within this area (see Appendix 1). These results allow us to use this model in this configuration for the investigation of climate response to the scenario of urban expansion being considered.

Table 1. Observed mean rural temperature (T), observed mean summer precipitation sum (P), averaged over weather stations located within domain D3, and statistics of comparison between modelled and observed values for T, P, and UHII: mean (ME) and root-mean square (RMSE) errors and correlation coefficients (R).

Year	Mean rural T, °C	Mean P, mm	Statistics of comparison between modelled and observed values										
			Mean rural T			UHII (Balchug st.)			UHII (MSU st.)			Mean P	
			ME, °C	RMSE, °C	R	ME, °C	RMSE, °C	R	ME, °C	RMSE, °C	R	ME, mm	ME, %
2009	16.7	215	0.2	1.4	0.96	0.2	1.3	0.73	0.4	1.2	0.55	55	25.4
2010	21.5	158	0.1	1.6	0.97	0.1	1.4	0.78	0.4	1.5	0.56	6	4.0
2013	18.2	240	0.0	1.2	0.97	0.1	1.3	0.77	0.2	1.3	0.63	28	11.8
2014	18.0	144	0.3	1.4	0.97	0.0	1.3	0.83	0.4	1.3	0.75	7	4.9
2015	17.1	197	-0.1	1.3	0.96	0.2	1.2	0.81	0.1	1.1	0.71	2	1.2

3.2. Model response to city expansion in terms of temperature and precipitation

The model response to the expansion of the city was considered as the temperature difference between URBx2 and URB experiments for temperature and precipitation. To compare it with the meteorological anomalies created by the existing city, the model response to the emergence of modern city (switching on the UCM) was also estimated in the same way as the difference between URB and noURB experiments.

The response in the air temperature field shows that the city expansion to the suburbs not only leads to a significant local warming (more than 1°C) in these areas, but also has a heating effect on the surrounding areas, including the old city core and the city center. During the summer season, the average daily temperature within most of the city increased by 0.2-0.4°C (Figure 3a), the average night temperature (measured at 0 UTC) is 0.5-0.6°C. At first glance, these values are small enough. However, they correspond to an increase in an average UHI intensity for the city center by about 10%, and even more for the other parts of Moscow. For comparison, during 1977-2017, due to a combined effect of various factors, the UHI intensity for the city center increased by 0.7°C, and for the MSU campus region by 0.4°C. Therefore, these values are comparable.

The largest values of the temperature response to the city expansion in its center are observed in the evening, night, and early morning (Figure 4a), their maximum values exceed 1°C and are observed under conditions of anticyclonic weather. It is interesting that the daily course of such response is shifted forward by several hours in comparison with the daily course of UHII for a modern city (Figure 4c). This could be caused by the remoteness of the areas of city expansion from the center of Moscow. In addition, the dependence of the temperature response to the city expansion on rural temperature values (Figure 4b) and a similar dependence for modern-city UHII (Figure 4d) were clearly identified. The highest values of modern-city UHII and temperature response to the city expansion are observed under conditions of hot weather, which agrees with the results given for some other cities in [42–44]. This emphasizes the importance of such dependence in the context of the urban effects on the conditions of meteorological comfort, because human health study is most important in hot weather conditions, when relatively small changes in meteorological conditions can make a significant impact on people's health and well-being [5].

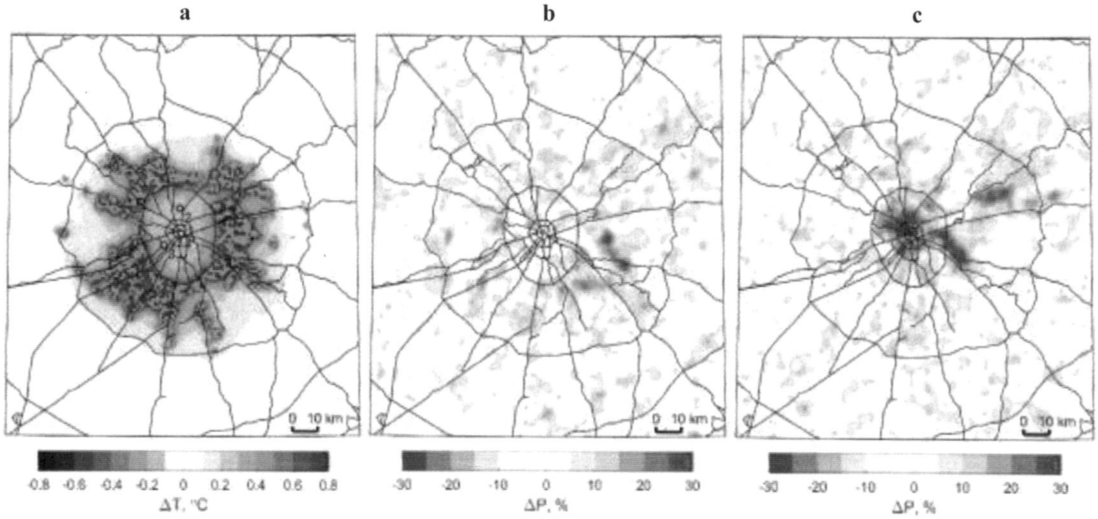

Figure 3. Spatial distribution of model response to city expansion in the field of mean air temperature ($\Delta T_{URBx2} = T_{URBx2} - T_{URB}$) averaged over considered summer seasons (a); relative model response to city expansion in the field of precipitation sum ($\Delta P_{URBx2} = 100\% \cdot (P_{URBx2} - P_{URB})/P_{URB}$) (b); relative model response to switching on the UCM in the field of precipitation sum ($\Delta P_{URB} = 100\% \cdot (P_{URB} - P_{noURB})/P_{noURB}$) (c). Model cells changed by the city expansion are shown by gray color in (a), and blue markers show the location of the main urban weater stations used in this study (1 – Balchug, 2 – VVC, 3 – MSU).

The modeling results also show that relatively small temperature changes in the central part of the city due urbanization at its periphery under conditions of extreme heat may significantly change the meteorological indices important for human thermal comfort assessment. For example, we analyzed this effect using the number of so-called "hot nights" (days with minimum temperature higher than 24 °C), which is a popular index of heat stress used in some studies, e.g. in [45, 46].

During the hot summer of 2010, the number of "hot nights" according URBx2 experiment was significantly higher that according to URB experiment: it increased by 3 for the Balchug station area (21 for URBx2 vs. 18 for URB), by 6 for VVC (10 vs. 4), and by 4 for MSU (8 instead of 4). It should be noted that for Balchug these numbers were calculated for the temperature representing the urban tile within the corresponding model cell, and for MSU and VVC the temperature for natural tile was used. In reality, according to the observational data for Balchug, VVC, and MSU, 25, 3, and 4 "hot nights" were observed, which is similar to URB experiment. And according to noURB experiment, "hot nights" were not observed at all, as well as at rural stations according to observations.

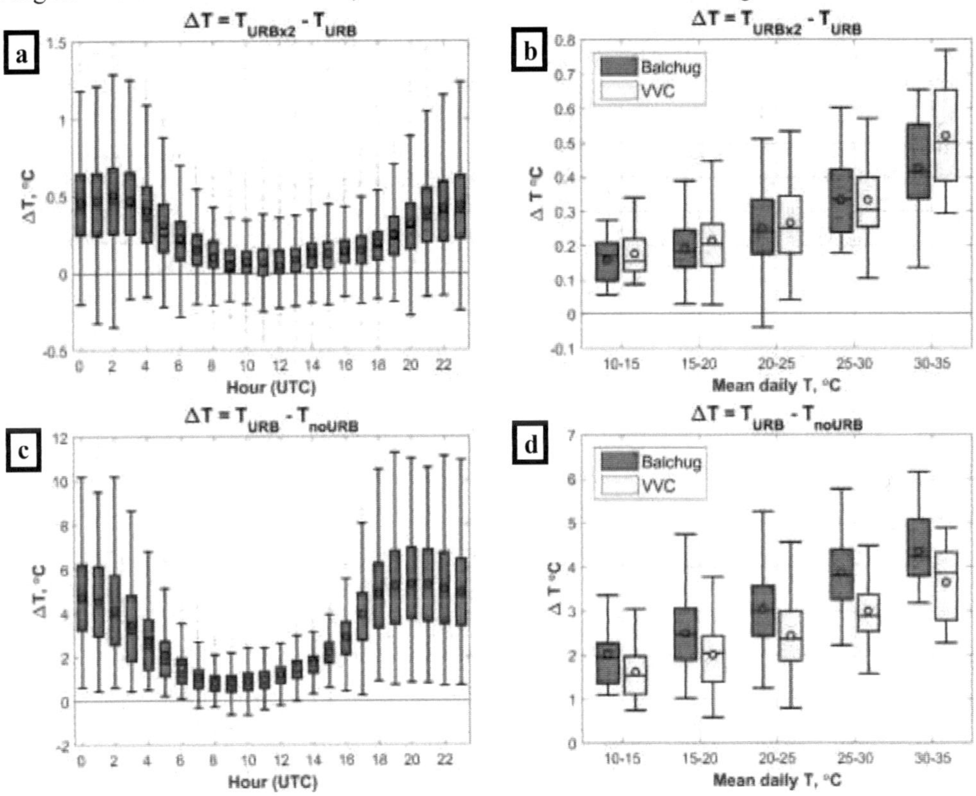

Figure 4. Whiskers diagrams representing the dependence of modelled temperature response to urban expansion (temperature difference between URBx2 and URB experiments) (a, b) to switching on the UCM for modern city (temperature difference between URB and noURB experiments) (c, d) the time of the day (a, c) and mean rural temperature (b, d). Diagrams (a, c) are constructed for hourly temperature values in the city center (Balchug station surroundings) and (b, d) for daily-mean temperature values in the city center (Balchug) and urban park (VVC station surroundings) for considered summer seasons.

A positive response to the city expansion is identified in the field of summer precipitation sum (Figure 3c), as well as an urban-caused precipitation anomaly for the modern city (Figure 3d). In both cases fields of the difference between the precipitation sums for the corresponding pairs of experiments (URBx2 – URB and URB – noURB) are rather noisy and spotty, which could be explained by the stochastic nature of summer convective showers. The spotting is even higher if we consider the same responses for certain summer season, and it decreases with increasing of the averaging period. 5-year averaging period is sufficient for clear identification of a positive urban-caused precipitation anomaly of the modern city located over its center and to the east of the city (Figure 3c) on its leeward site in relation to the prevailing western wind. This pattern agrees with the results of a number of investigations covering the problem of urban effect on precipitation [3, 47, 48].

The precipitation response to the scenario of urban expansion being considered is not so intensive, but it could be identified over a wider area (Figure 3c). Its maximum values are observed outside the "old" part of the city and it is also more intensive in the eastern (leeward) part of the region. If we consider the mean precipitation sum over the whole area of "final" domain D3 (180x180 km), the effect of the city expansion is approximately equal to the urban-caused anomaly for the modern city: in the both cases the precipitation sum increases approximately by 1.5% (see ΔP_1 values in Table 2). At the same time, the urban-caused precipitation anomaly for the modern city is more intensive within 20 km from its center, its mean value over this area is 10%, and in the city center and to the east of the city it reaches 30% (Figure 3c). The urban expansion scenario causes an approximately equal increase of precipitation over the "old" part of the city (within 20 km from its center) and over the area of new development lying mostly within 20-40 km from the center, for both of these areas the mean precipitation sum increases approximately by 5 % (see ΔP_2 and ΔP_3 values in Table 2). It should be noted that the inter-annual variability of these responses is rather high. However, for each of the 5 considered summer seasons they are positive for each of the three considered areas (Table 2).

Table 2. Mean values of the summer precipitation sum for the whole domain D3 (P_1), for the area within 20 km from the city center (P_2), and for the area within 20-40 km from the city center (P_3), and values of the model response in the field of summer precipitation to switching on the UCM for modern city (ΔP_{URB}) and to the city expansion (ΔP_{URBx2}) averaged over listed areas (ΔP_1, ΔP_2, ΔP_3).

Year	P_{URB}, mm			$\Delta P_{URB} = P_{URB} - P_{noURB}$						$\Delta P_{URBx2} = P_{URBx2} - P_{URB}$					
	P_1	P_2	P_3	ΔP_1		$\Delta P2$		$\Delta P3$		ΔP_1		$\Delta P2$		$\Delta P3$	
				mm	%	mm	%	mm	%	mm	%	mm	%	mm	%
2009	270	297	289	3	*1.0*	18	*6.3*	9	*3.2*	3	*1.0*	20	*6.8*	12	*4.3*
2010	164	175	177	2	*1.5*	14	*8.4*	5	*2.9*	2	*1.4*	4	*2.2*	7	*4.0*
2013	268	263	267	8	*2.9*	71	*27.1*	15	*5.6*	9	*3.2*	5	*1.5*	29	*10.3*
2014	151	154	164	3	*1.8*	19	*12.2*	6	*3.5*	3	*1.6*	10	*5.7*	7	*4.0*
2015	200	191	202	1	*0.5*	1	*0.5*	6	*2.9*	2	*0.9*	10	*5.5*	6	*2.9*
Mean	211	216	220	3	*1.5*	24	*10.9*	8	*3.6*	4	*1.6*	10	*4.3*	12	*5.1*

4. Conclusions

The main result of this study is the demonstration of possibilities to simulate the meteorological regime over large urban agglomerations with high spatial resolution within the framework of regional climate modelling and its application to the investigation of climatic responses to various urban developments. The numerical experiments with such model (COSMO-CLM) adapted to the conditions of Moscow region, supplemented with TERRA_URB urban canopy model, and supplied with realistic parameters of urban surface show its ability to successfully reproduce the summer meteorological regime of the research area, including the features of Moscow UHI.

The scenario of doubling of the city size by its expansion on the periphery was considered as a testing case for the investigation using this modelling approach to evaluate the climatic effect of urban developments. The simulations for this scenario show that it leads not only to a local temperature increase in the area of new developments, but also provides a heating effect on the central part of the city and increases the UHII in the city center by 10% or higher (the average daily temperature increases by 0.3 ° C, and the average nocturnal temperatures, by 0.6 ° C). This result shows the good agreement with the urban amplification of global warming diagnosed for Moscow according to meteorological observations over the past 40 years [14] and partly explains its causes. At the same time, the temperature response to the city expansion is maximal in conditions of extreme heat and could impair the meteorological comfort and cause increased heat stress, which is shown by the example of the frequency of "hot nights" events during the summer of 2010.

In addition, the simulation results show that Moscow can be characterized by a positive urban-caused anomaly of total summer precipitation (in the city and at its leeward side it increased by 10-30% by the influence of modern city). The scenario of city expansion reinforces this anomaly by 5-10% and enlarges its area. Understanding the physical mechanisms of this anomaly formation, as well as its more detailed characteristics (such as its dependence on the precipitation type and intensity) require more detailed studies. However, the fact of its existence and sensitivity to urban expansion is an important result.

Appendix 1

The mean rural temperature was calculated according to observations at 9 stations located within 110 km from Moscow and surrounding it: Naro-Fomisk (WMO id 27611), Novo-Ierusalim (27511), Klin (27417), Dmitrov (27419), Aleksandrov (27428), Pavlovskij Posad (27523), Kolomna (27625), Serpuhov (27618), and Malojaroslavec (27606). The urban climate features were characterized according to observations at Balchug station (27605), VVC (27612), and the meteorological observatory of Lomonosov Moscow State University (MSU). To calculate the mean precipitation sum over Moscow region, in addition to the listed stations we used observations at Podmoskovnaja station (27518), Nemchinovka (27515), and Tushino (27619). To verify the ability of the model to reproduce the features of spatial distribution of the temperature in Moscow and its surroundings (Figure 2d-e), in addition to all these weather stations we used observations at Moscow airports, new AWSs, and the network of AAQSs.

Appendix 2

The key differences between the tuned model configuration and the default configuration are: 1) using new parameterization of evaporation from bare soil [49] instead of [50]; 2) switching on a new vegetation canopy parameterization [51,52]; 3) using newer aerosol climatology [53] instead of [54]; 4) a significant decrease (5 times for D1/D2 and 10 times for D3) of the minimum possible values of the turbulent diffusion coefficients and values of the scale of subgrid thermal inhomogeneity affecting the generation of turbulent kinetic energy according to [26–28], 5) changing the type of the root distribution in soil and increasing the rooting depth 2.1 times; 6) increasing the minimal possible value of the leaf stomatal resistance by 1/3; 7) increasing the soil hydraulic conductivity 1.62 times according to [55]; 8) decreasing the parameter affecting the subgrid cloudiness to decrease its amount, which is overestimated by the model in its default configuration [55–57].

Acknowledgements

The bulk of the research (supercomputer modeling, data management and analysis) performed by M.I. Varentsov and P.I. Konstantinov was supported by the grant program of Russian Science Foundation (project no. 17-77-20070 "An initial assessment and projection of the bioclimatic comfort in Russian cities in XXI century against the context of climate change"). The human thermal comfort assessment in Moscow (performed by P.I. Konstantinov) was supported by the Russian Foundation for Basic Research (grant no.15-35-70006-mol_a_mos).

References

[1] Oke T R 1987 *Boundary layer climates* (London: Routledge)
[2] Landberg G E 1981 *Urban climate* (New York: Academic press)
[3] Han J Y, Baik J J and Lee H 2014 Urban impacts on precipitation *Asia-Pacific J. Atmos. Sci.* **50** 17–30

[4] Lemonsu A and Masson V. 2002 Simulation of a summer urban breeze over Paris *Bound. Layer Meteorol.* **104** 463–90
[5] Konstantinov P I, Varentsov M I and Malinina E P 2014 Modeling of thermal comfort conditions inside the urban boundary layer during Moscow's 2010 summer heat wave (case-study) *Urban Clim.* **10(P3)** 563–72
[6] Cox W 2017 *Demographia World Urban Areas (World Agglomerations) 13th Annual edition* (Belleville, Illinois: Wendel Cox Consultancy)
[7] Lokoshchenko M A 2014 Urban "heat island" in Moscow *Urban Clim.* **10 (P3)** 550–62
[8] Lee D O 1992 Urban warming?—an analysis of recent trends in London's heat island *Weather* **47** 50–6
[9] Wilby R L 2003 Past and projected trends in London's urban heat island *Weather* **58** 251–60
[10] Liu W, Ji C, Zhong J, Jiang X and Zheng Z 2007 Temporal characteristics of the Beijing urban heat island *Theor. Appl. Climatol.* **87** 213–21
[11] Liu W, You H and Dou J. 2009 Urban-rural humidity and temperature differences in the Beijing area. *Theor. Appl. Climatol.* **96** 201–7
[12] Longxun C, Wenqin Z, Xiuji Z and Zijiang Z 2003 Characteristics of the heat island effect in Shanghai and its possible mechanism. *Adv. Atmos. Sci.* **20** 991–1001
[13] Kislov A V and Konstantinov P I 2011 Detailed spatial modeling of temperature in Moscow *Russ. Meteorol. Hydrol.* **36** 300–6
[14] Kislov A V., Varentsov M I, Gorlach I A and Alekseeva L I 2017 "Heat island" of the Moscow agglomeration and the urban-induced amplification of global warming *Moscow Univ. Vestnik Ser. 5 Geogr.* **4** 12–9
[15] Buechley R W, Van Bruggen J and Truppi L E 1972 Heat island = death island? *Environ. Res.* **5** 85–92
[16] Tan J *et al* 2010 The urban heat island and its impact on heat waves and human health in Shanghai. *Int.J. Biometeorol.* **54** 75–84
[17] Shaposhnikov D *et al* 2014 Mortality related to air pollution with the Moscow heat wave and wildfire of 2010 *Epidemiology* **25** 359–64
[18] Böhm U, Kücken M, Ahrens W, Block A, Hauffe D, Keuler K, Rockel B and Will A 2006 CLM— the climate version of LM: brief description and long-term applications *COSMO Newsl.* **6** 225–35
[19] Dee D P *et al* 2011 The ERA-Interim reanalysis: Configuration and performance of the data assimilation system *Q.J. R. Meteorol. Soc.* **137** 553–97
[20] von Storch H, Langenberg H and Feser F 2000 A Spectral Nudging Technique for Dynamical Downscaling Purposes *Mon. Weather Rev.* **128** 3664–73
[21] Varentsov M I, Verezemskaya P S, Zabolotskikh E V and Repina I A 2016 Evaluation of the quality of polar low reconstruction using reanalysis and regional climate modelling *Sovremennye Problemy Distantsionnogo Zondirovaniya Zemli iz Kosmosa* **4** 168–91
[22] Feser F and Barcikowska M 2012 The influence of spectral nudging on typhoon formation in regional climate models *Environ. Res. Lett.* **7** 14024-34
[23] Wyngaard J C 2004 Toward Numerical Modeling in the "Terra Incognita" *J. Atmos. Sci.* **61** 1816–26
[24] Tiedtke M 1989 A comprehensive mass flux scheme for cumulus parameterization in large-scale models *Mon. Weather Rev.* **117** 1179–800
[25] Bontemps S, Defourny P, Bogaert E, Arino O, Kalogirou V and Perez J R 2011 GLOBCOVER 2009 Products Description and Validation Report *ESA Bull.* 136:53
[26] Cerenzia I, Tampieri F and Tesini M S 2014 Diagnosis of Turbulence Schema in Stable Atmospheric Conditions and Sensitivity Tests *COSMO Newsl.* **14** 28–36
[27] Rossa A M, Domenichini F and Szintai B 2012 Selected COSMO-2 verification results over North-eastern Italian Veneto *COSMO Newsl.* **12** 64–71

[28] Buzzi M, Rotach M W, Raschendorfer M and Holtslag A A M 2011 Evaluation of the COSMO-SC turbulence scheme in a shear-driven stable boundary layer *Meteorol Zeitschrift.* **20** 335–50
[29] Kislov A V, Varentsov M I and Tarasova L L 2015 Role of spring soil moisture in the formation of large-scale droughts in the East European Plain in 2002 and 2010 *Izv. Atmos. Ocean. Phys.* **51** 405–11
[30] Grumm R H 2011 The central European and Russian heat event of July-August 2010 *Bull. Am. Meteorol. Soc.* **92** 1285–96
[31] Wouters H, Demuzere M, Ridder K D and Van Lipzig N P M 2015 The impact of impervious water-storage parametrization on urban climate modelling *Urban Clim.* **11** 24–50
[32] Wouters H, Demuzere M, Blahak U, Fortuniak K, Maiheu B, Camps J, Tielemans D and van Lipzig N P M 2016 Efficient urban canopy parametrization for atmospheric modelling: description and application with the COSMO-CLM model for a Belgian Summer *Geosci. Model. Dev.* **9** 3027–54
[33] Nunez M and Oke T R 1977 The Energy Balance of an Urban Canyon *J. Appl. Meteor.* **16** 11–9
[34] Flanner M G 2009 Integrating anthropogenic heat flux with global climate models *Geophys. Res. Lett.* **36** L02801
[35] Samsonov T E, Konstantinov P I and Varentsov M I 2015 Object-oriented approach to urban canyon analysis and its applications in meteorological modeling. *Urban Clim.* **13** 122–39
[36] Stewart I D and Kennedy C A 2017 Metabolic heat production by human and animal populations in cities *Int. J. Biometeorol.* **61** 1159–71
[37] Ginzburg A S, Belova I N, Raspletina N V 2011 Anthropogenic heat fluxes in urban agglomerations *Dokl. Earth Sci.* **439** 256–9
[38] Aleksandrov G G, Belova I N and Ginzburg A S 2014 Anthropogenic heat flows in the capital agglomerations of Russia and China *Dokl. Earth Sci.* **457** 101–6
[39] Varentsov M I, Samsonov T E, Kislov A V and Konstantinov P I 2017 Simulations of Moscow agglomeration heat island within framework of regional climate model COSMO-CLM *Moscow Univ. Vestnik Ser. 5 Geogr.* **6**. In print.
[40] Gorchakov G I, Kadygrov E N, Kunitsyn V E, Zakharov V I, Semutnikova E G, Karpov A V, Kurbatov G A, Miller E A and Sitanskii S I 2014 The Moscow heat island in the blocking anticyclone during summer 2010 *Dokl. Earth Sci.* **456** 736–40
[41] Kramar V F, Baykova E, Kallistratova M, Kouznetsov R and Kulichkov S 2014 Ground-based remote sensing of the ABL structure in Moscow and its use to estimate pollutant surface emission rates *J. Appl. Meteorol. Climatol.* **53** 1272–81
[42] De Ridder K, Maiheu B, Lauwaet D, Daglis I, Keramitsoglou I, Kourtidis K, Manunta P and Paganini M 2016 Urban Heat Island Intensification during Hot Spells - The Case of Paris during the Summer of 2003 *Urban Sci.* **1**
[43] Li D and Bou-Zeid E 2013 Synergistic interactions between urban heat islands and heat waves: The impact in cities is larger than the sum of its parts *J. Appl. Meteorol. Climatol.* **52** 2051–64
[44] Schatz J and Kucharik C J 2015 Urban climate effects on extreme temperatures in Madison, Wisconsin, USA. *Environ Res. Lett.* **10** 94024
[45] Macnee R G D and Tokai A 2016 Heat wave vulnerability and exposure mapping for Osaka City, Japan. *Environ. Syst. Decis.* **36** 368–76
[46] Nicholls N, Skinner C, Loughnan M and Tapper N 2008 A simple heat alert system for Melbourne, Australia *Int. J. Biometeorol.* **52** 375–84
[47] Bornstein R and Lin Q 2000 Urban heat islands and summertime convective thunderstorms in Atlanta: Three case studies *Atmos. Environ.* **34** 507–16
[48] Shepherd J M 2006 Evidence of urban-induced precipitation variability in arid climate regimes *J. Arid Environ.* **67** 607–28

[49] Schulz J-P and Vogel G 2016 An evaluation of the simulated bare soil evaporation of an atmospheric model *Geophys. Res. Abstr.* **18** 8559
[50] Manabe S 1969 Climate and the ocean circulation: 1. The atmospheric circulation and the hydrology of the Earths' surface *Mon. Weather. Rev.* **97** 739–805
[51] Viterbo P and Beljaars A C M 1995 An improved land surface parameterization scheme in the ECMWF model and its validation *J. Clim.* **8** 2716–48
[52] Schulz J-P and Vogel G 2017 An improved representation of the land surface temperature including the effects of vegetation in the COSMO model *Geophys. Res. Abstr.* **19** 7896
[53] Tegen I, Hollrig P, Chin M, Fung I, Jacob D and Penner J 1997 Contribution of different aerosol species to the global aerosol extinction optical thickness: Estimates from model results *J. Geophys. Res.* **102** 23895–915
[54] Tanré D, Geleyn J F and Slingo J M 1984 First results of the introduction of an advanced aerosol-radiation interaction in the ECMWF low resolution global model. *Aerosols their Clim. Eff.* 133–77
[55] Ho-Hagemann H T M, Gröger M, Rockel B, Zahn M, Geyer B and Meier H E M 2017 Effects of air-sea coupling over the North Sea and the Baltic Sea on simulated summer precipitation over Central Europe *Clim. Dyn.* 1–26
[56] Jaeger E B, Anders I, Lüthi D, Rockel B, Schär C and Seneviratne S I 2008 Analysis of ERA40-driven CLM simulations for Europe *Meteorol. Zeitschrift* **17** 349–67
[57] Pfeifroth U, Hollmann and Ahrens B 2012 Cloud cover diurnal cycles in satellite data and regional climate model simulations *Meteorol. Zeitschrift* **21** 551–60

Statistical analysis of experimental data for mathematical modeling of physical processes in the atmosphere

P A Karpushin[1,3], Yu B Popov[2], A I Popova[2], K Yu Popova[1], N P Krasnenko[1,3] and A V Lavrinenko[4]

[1]Tomsk State University of Control Systems and Radioelectronics, Russia, Tomsk,
[2]Surgut State University, Russia, Surgut,
[3]Institute of Monitoring of Climatic and Ecological Systems SB RAS, Russia, Tomsk,
[4]V.E. Zuev Institute of Atmospheric Optics SB RAS, Russia, Tomsk,

E-mail: pavel.a.karpushin@tusur.ru, kseniia.i.popova@tusur.ru, krasnenko@mces.ru

Abstract. In this paper, the probabilities of faultless operation of aerologic stations are analyzed, the hypothesis of normality of the empirical data required for using the Kalman filter algorithms is tested, and the spatial correlation functions of distributions of meteorological parameters are determined. The results of a statistical analysis of two-term (0, 12 GMT) radiosonde observations of the temperature and wind velocity components at some preset altitude ranges in the troposphere in 2001–2016 are presented. These data can be used in mathematical modeling of physical processes in the atmosphere.

In the last few years, a dynamic stochastic approach based on the theory of Kalman filtering has been widely used in mathematical modeling of physical processes in the atmosphere and for the development and study of algorithms of reconstruction of mesometeorological fields. The use of the Kalman filter for solving these problems requires testing for Gaussian distribution of the sought-for meteorological parameters [1]. Therefore, we performed corresponding investigations, in which long-term aerologic observations in the troposphere, given in standard bulletins and telegrams of the World Meteorological Organization (WMO) for 2001–2016 in the KN-04 code, were used as empirical data. In the present work, the hypothesis of normality of the empirical data distributions is tested, the spatial correlation functions of the meteorological parameters are calculated, and the probabilities of faultless operation of the observation channels are analyzed.

The data of aerologic stations were grouped according to their geographical coordinates into typical mesometeorological experimental sites. In the first stage of the initial aerologic data processing, the code combinations of telegrams were analyzed and the coding errors were corrected. Using a linear interpolation, the vertical profiles of the aerologic data of each of the chosen stations of the Russian Federation were reduced to a system of standard isobaric altitude levels and geometric altitudes [2, 3].

With the data array obtained, the aprioristic probabilities of faultless operation of the measurement channels were analyzed, and the allowable threshold values were determined to decide on the presence or absence of 1) useful information in the measurement channels, 2) anomalous data overshoots in the measurements, 3) an atmospheric front within the mesoscale experimental site being considered.

Content from this work may be used under the terms of the Creative Commons Attribution 3.0 licence. Any further distribution of this work must maintain attribution to the author(s) and the title of the work, journal citation and DOI.
Published under licence by IOP Publishing Ltd

To estimate the probability of faultless operation q and the probability of measurement channel failure $(1 - q)$, data arrays were compiled for each station from the entire aerologic data archive for a fixed period of time. The data arrays were formed for two crucial months: January and July. A preliminary analysis of the meteorological data on the presence of data omissions demonstrated that the aprioristic probability of faultless operation was 0.89 for both winter and summer.

In our analysis, the presence in the observation data of anomalous errors arising in the course of compiling, re-writing, and deciphering of the standard meteorological telegrams was taken into account. The presence of these inhomogeneities may lead to an increase in the errors and to failures in the operation of forecast algorithms. The analysis of the calculated statistical characteristics allowed us to reveal the anomalous errors. The difference between the verified value and the arithmetic average was used for this purpose. According to the theory of hypothesis testing, the modulus of this difference was compared with $N\sigma$, where σ is the standard deviation of the chosen meteorological parameter [4] and $N = 3$.

An important point in the analysis of the aerologic data is the detection of the passage of an atmospheric front. For this purpose, a comparison of the differences between the increments of pressure, temperature, and wind direction at two subsequent kth and (k+1)th time moments were used. A sharp change in the wind direction that breaks the monotonic behavior of the meteorological parameters is an obvious indication of the passage of an atmospheric front. The aprioristic probabilities of front detection and the threshold values in decision-making on the absence or presence of a front were calculated for the meteorological measurements at each of the stations.

To test the statistical hypothesis of normality of the empirical meteorological parameters being analyzed, the coefficients of asymmetry (A) and excess (E) were analyzed together with numerical values of the Kolmogorov–Smirnov normality test criterion [4, 5, 6]. To estimate the deviation of the curve of a normal distribution from the curves of empirical distributions, formulas [7]

$$\sigma_A \cong \sqrt{\frac{6}{n+4}}, \qquad \sigma_E \cong \sqrt{\frac{24*(n-5)}{n*(n+7)}} \tag{1}$$

were used, where σ_A is the standard root-mean-square error of the asymmetry coefficient, σ_E is the same error of the excess coefficient, and n is the number of examined realizations. The series was considered to be normally distributed, if the conditions

$$|A| \leq 3\sigma_A, \ |E| \leq 3\sigma_E \tag{2}$$

were satisfied. During our analysis, we investigated some temporal statistical characteristics of the meteorological fields. To calculate the statistical characteristics, the following expressions were used:

– meteorological parameters averaged over an ensemble of realizations or the first initial moments,

– variance or the second-order central moment of a random process,

– excess coefficient (characterizes the position of the maximum in the curve of an empirical distribution relative to the curve of a normal distribution),

– asymmetry coefficient (characterizes the skewness of the distribution curve).

The results of our investigations demonstrated an insignificant deviation of the empirical distribution functions of the observation data from a normal Gaussian distribution law. Inequality (2) holds true for each station of the experimental site. The insignificant deviations from the normal distribution law are explained by the presence of systematic and random errors in the data of observations of meteorological fields in the atmosphere.

The statistical analysis demonstrated that the empirical distributions of the examined meteorological parameters can be described with sufficient accuracy by ae normal distribution, which is a major condition for the use of the Kalman filter in the algorithms for the reconstruction of mesometeorological fields. This distribution is a good mathematical model for the observable random variables.

The correlation functions for the stations were calculated by the standard formulas [8]:

$$\mu_{\xi}(\rho) = \frac{1}{n-1}\sum_{i=1}^{n}\left[\xi(r_i) - m_{\xi}(r_i)\right]\left[\xi(r_i+\rho) - m_{\xi}(r_i+\rho)\right], \quad (3)$$

where $\xi(r_j)$ are the values of the meteorological parameter at points r_j, respectively, $m_{\xi}(\rho)$ is the mathematical expectation of the meteorological parameter, and ρ is the distance. Figures 1 and 2 show the spatial correlation functions for the chosen experimental site, indicated altitudes, and January and July periods.

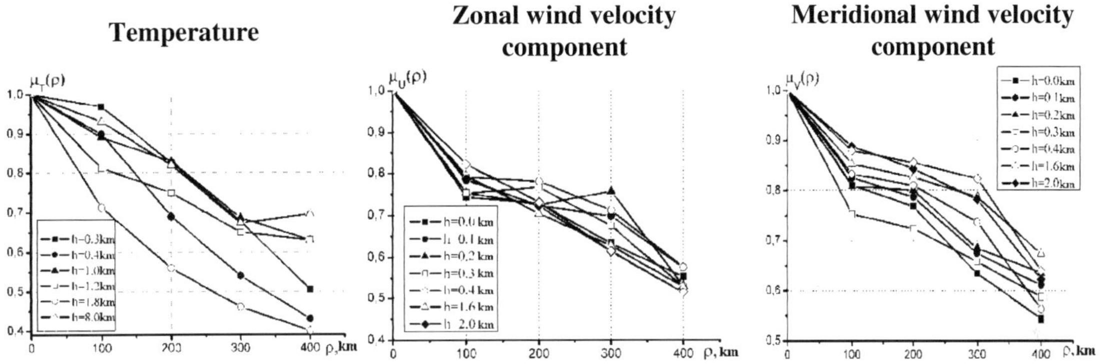

Figure 1. Spatial empirical correlation functions of the meteorological parameters, January.

The spatial correlations of the temperature and wind velocity components at all examined altitudes (Figure 1) decrease with increasing distance ρ.

In summer, the spatial correlations of the temperature and wind velocity components at all examined altitudes shown in Figure 2 behave similarly to those shown in Figure 1.

The correlation functions shown in Figs. 1 and 2 allow using the analytical expression

$$\mu(\rho) = e^{-\beta \cdot \Delta\rho} \quad (4)$$

as an approximating function. In spite of the fact that the quality of approximation is worse than that in [4, 9], it allows one to simplify significantly the Kalman filter algorithm of spatial forecasting.

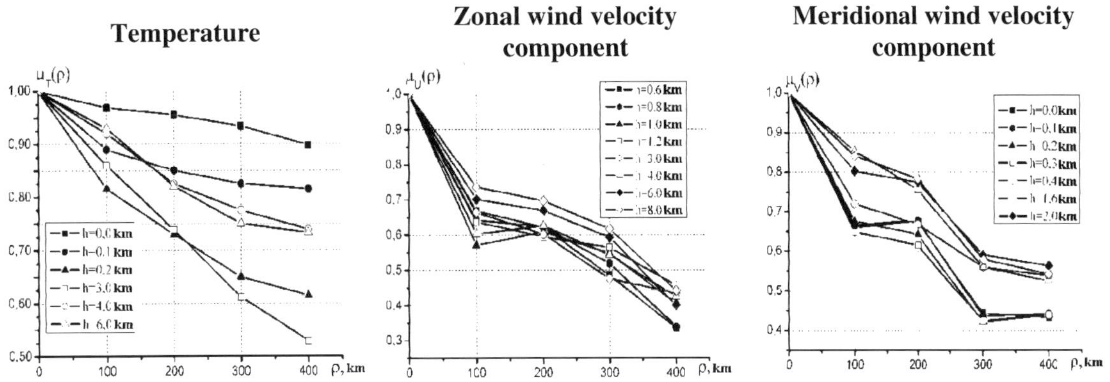

Figure 2. Spatial empirical correlation functions of the meteorological parameters, July.

The results were obtained with financial support of the Ministry of Education and Science of the Russian Federation (project no. 5.3279.2017/PSN) and of the Program of Basic Research of the Siberian Branch of the Russian Academy of Sciences (project no. IX.138.2.5).

References

[1] Popov Yu B 2010 Estimation of the parameters of the atmospheric state using four-dimensional dynamic stochastic model and the linear Kalman filter Part 1 Basic Methodology *TUSUR Rep.* **21** 5-11

[2] Lavrinenko A V 2006 *Multidimensional dynamic stochastic models and their use in applied problems. Candidate's Dissertation* (Tomsk: IAO SB RAS) p 136

[3] Popova A A 2006 *Application of adaptive algorithms in numerical method of spatiotemporal extrapolation of mesometeorological fields. Candidate's Dissertation* (Surgut: IAO SurSU) p 147

[4] Il'in S N 2005 *Dynamic stochastic method of spatial extrapolation of meteorological fields in a mesoscale region Dissertation* (Tomsk: IAO) p 120

[5] Kobysheva N V and Narovlyanskii G Ya 1978 *Climatic Processing of Meteorological Information* (Leningrad: Gidrometeoizdat) p 295

[6] Manita A D 2001 *Probability Theory and Mathematical Statistics: A Textbook.* (Moscow: Publishing Department of Educational and Scientific Center of Pre-University Education) p 120

[7] Bol'shakov V D 1983 *Theory of Observational Errors: A Textbook for High Schools – 2nd Addition Rev. Suppl.* (Moscow: Nedra) p 223.

[8]. Popov Yu B and Popova A A *2008 Optimal Filtration and its Application to Problems of Monitoring of the Parameters of the Atmospheric State within Local Territories* (Khanty-Mansiysk: Poligrafist) p 188

[9] Komarov V S, Popov Yu B, Suvorov S S and Kurakov V A 2004 *Dynamic Stochastic Methods and Their Use in Applied Meteorology* (Tomsk: Publishing House of the IAO SB RAS) p 236

Numerical modelling of the Earth's ionosphere F region

P A Ostanin[1], **D V Kulyamin**[2], **V P Dymnikov**[3]

[1] Moscow Institute of Physics and Technology (MIPT SU), Institutskiy per., 9, Dolgoprudny, Moscow Region, 141701, Russian Federation

[2] Lomonosov Moscow State University, Research Computing Center (RCC), Leninskye gory, 1, Moscow, 118991, Russian Federation

[3] Institute of Numerical Mathematics of the Russian Academy of Sciences (INM RAS) Gubkin str., 8, Moscow, 118333, Russia

E-mail: ostanin.pavel@phystech.edu

Abstract. This paper presents the first version of a new INM RAS Earth's ionosphere F region dynamical model. A complete set of model equations is formulated taking into account all the key physical processes that form the global state of the ionospheric F region (plasma chemistry, ambipolar diffusion, wind transport, drift across magnetic lines). For the numerical solution, a splitting method based on the physical processes and geometric directions is proposed. The first stage of splitting in a quasi-two-dimensional approximation setting with a projection of ambipolar diffusion on the vertical direction is considered. It is numerically implemented stepwise using various difference schemes for three separate model formulations (taking into account diffusion only along the vertical direction, considering a realistic direction of diffusion along the magnetic field excluding and including a mixed derivative term). The applicability, efficiency, conservation, and monotonicity of these numerical methods are analyzed. The first numerical experiments show convergence of the numerical solution to a stationary vertical profile specific to the F region. The greatest consistency with the observed profiles is obtained in the mid-latitudes. Using the thus constructed model it is shown that the electron density profile is most sensitive to the neutral temperature and ionization level with qualitatively different structures of the corresponding modes of variability.

1. Introduction

This paper addresses the fundamental problem of describing and predicting the mean global state and variability of the Earth's ionosphere, as well as the physical mechanism of its formation, by numerical modeling of the F region with particular attention to the construction of efficient numerical methods and algorithms of their implementation. This work lies within the framework of a wider research direction currently being carried out at the Institute of Numerical Mathematics of the Russian Academy of Sciences (INM RAS), which is connected with the construction of upper atmosphere global models as parts of a comprehensive unified INM RAS Earth's system model. It aims at the development of a new consistent high-level coupled ionosphere-thermosphere dynamical numerical model.

The main objective of this paper is to present a numerical implementation and first results of numerical experiments with a new INM RAS Earth's ionosphere dynamical model. Primary attention is paid to an analysis of the model equations and a detailed description of the

Content from this work may be used under the terms of the Creative Commons Attribution 3.0 licence. Any further distribution of this work must maintain attribution to the author(s) and the title of the work, journal citation and DOI.
Published under licence by IOP Publishing Ltd

numerical methods with investigation of their properties and efficiency. Further perspective is the integration of this model into a developed INM RAS thermosphere global dynamical model [1] in order to create the first version of a consistent upper atmosphere coupled model and couple it with INM RAS models of lower atmosphere and ionosphere layers [2].

The importance of this problem is caused by recently increased practical interest in the research and forecasting of space weather, which is connected with the specific role of ionospheric conditions for global radio and satellite communication, as well as for the whole space industry. The state of thermosphere-ionosphere system determines the characteristics of the low-orbit satellite motion and the conditions for radio signal propagation (which ensures stable operation of distant communications and global navigation systems).

At the same time, there is a certain disconnection nowadays between the traditional semi-empirical approach to the upper atmosphere study and the high-tech numerical modeling methods successfully applied to weather and climate change forecasting. The problem of describing the Earth's ionosphere characteristics is traditionally solved by processing the available experimental data and the construction of empirical models. These models of upper atmosphere basically describe the climatological state and for the most part do not take into account the variability of the environment caused by both external and internal factors (such as significant perturbations of radiation and electromagnetic fields, nonlinear wave processes in the thermosphere and ionosphere, etc.). In the meantime, these models are widely used to solve important practical problems. The most known examples of such reference models are IRI, SIMP, GOST [3-5] and others.

It is also important to note that the upper atmosphere numerical models are smaller in number and less developed in comparison with the modern weather forecast and climate models which take into account the lower atmosphere. The most advanced ionospheric models are developed mainly at global national centers and consortiums of various institutions; examples are the advanced coupled models of the thermosphere and ionosphere, such as the large American National Center for Atmospheric Research (NCAR) models TIEGCM, TIEM-GCM [6,7]; British UCL CTIP-CMAT [8]; domestic models UAM [9], GCM-TIP [10]. Modern ionospheric models developed in the past few decades (such as GAIM (originally TDIM) [11], or a Russian CAO assimilative ionospheric model [12]) utilize approaches based on data assimilation systems (using GPS network data). Mathematically, a common disadvantage of today's upper atmospheric models is that their basic versions were developed in the 1980-1990s under severe restrictions associated with the relatively poor performance of the computer facilities at that time, as well as the insufficient efficiency of the numerical methods. These models essentially use the generalized features of the mean ionospheric structure (the exponential character of the neutral components distribution in the radial direction, the predominant propagation of charged particles along the magnetic field lines, etc.), and do not take into account the latest research data of the thermosphere and ionosphere. Thus, creating a substantially new methodology for modeling and predicting the global state and variability of the ionosphere-thermosphere system based on the successful experience of INM RAS in Earth's climate modeling, research, and numerical forecasting is very important.

In this work the first version of the developed INM RAS ionosphere F region global dynamical model is presented. A description of the complete problem statement, model physical foundations, and solved equations is contained in the second section of the paper. The third chapter is devoted to a description of a stepwise methodology for this model implementation and a detailed analysis of the numerical methods used. The fourth section presents the results of numerical experiments. The fifth chapter contains a final discussion of the results and plans for further work.

2. Ionosphere F region model

2.1. Problem formulation

The first version of the dynamical ionosphere model is based on the following assumptions: considering only the F region, photochemical dominance of atomic oxygen O ionization and O^+ ion recombination with major air components and, thereby, a single-ion assumption, plasma quasineutrality (equality of the electron and ion concentrations $n_e = n_i = n$), a simultaneous motion of the electrons and ions along magnetic field lines in ambipolar diffusion, electromagnetic drift dominance, a dipole Earth's magnetic field with coincidence of the magnetic and geographic poles. We considered an altitude region of 100 – 500 km, which also includes lower ionospheric layers (E region). However, in the first version of the dynamical ionosphere model the emphasis is on the formation of the F layer and the role of dynamic processes. In this case the physical processes of the lower layers are not reproduced by this model [13]. The lower boundary is formally taken far from the electron concentration maximum in order to reduce its impact on the calculated distribution. A detailed consideration of D and E ionosphere layers in a unified atmosphere INM RAS model is implemented in a separate module by virtue of the dominance of photochemistry processes in these regions.

The main equation of the model is the continuity equation for the electron concentration n:

$$\frac{\partial n}{\partial t} + \operatorname{div}(n\vec{u}) = P - kn, \tag{1}$$

where $\vec{u}$ is the plasma transfer velocity connected with diffusion, neutral wind and electromagnetic drift. The velocity can be calculated [13], and equation (1) can be rewritten in the following form:

$$\frac{\partial n}{\partial t} = P - kn - \operatorname{div}(n\vec{u}_{n\|}) - \operatorname{div}\left(n\frac{[\vec{E}\times\vec{B}]}{B^2}\right) + \operatorname{div}\left(D\left[\vec{\nabla}_{\|}n + n\frac{1}{T_p}\vec{\nabla}_{\|}T_p - \frac{nm_i}{2kT_p}\vec{g}_{\|}\right]\right). \tag{2}$$

In this equation $D = \dfrac{2kT_p}{m_i\nu_{in}}$ is the ambipolar diffusion coefficient, $\nu_{in} \approx 10^{-14} \cdot \sqrt{\frac{1}{2}(T_i + T_n)} \cdot$ $\cdot n_O$ [s^{-1}] is the frequency of O^+ collisions with neutrals, m_i is the ion atom mass, $\vec{g}$ is the gravitational field of the Earth, $\vec{E}$ is the external electric field (magnetospheric convection, etc.), $\vec{B}$ is the Earth's magnetic field, $\overrightarrow{u_n}$ is the neutral field velocity, T_n is the neutral temperature, $T_p = \frac{1}{2}(T_e + T_i)$ is the mean of the electron and ion temperatures, and the index $\|$ indicates projection on the magnetic field lines [13]. The components in the right hand side stand for the ionization process from O to O^+ and for the recombination loss. In this first version of the model only fixed daytime values of P and k are used: $P = 4 \cdot 10^{-7} \cdot n_O(z)$ [s^{-1}], $k = 1{,}2 \cdot 10^{-12} \cdot n_{N_2}(z) + 2{,}1 \cdot 10^{-11} \cdot n_{O_2}(z)$ [s^{-1}]. In the expressions above n_O, n_{O_2}, n_{N_2} are neutral concentrations, approximately calculated in accordance with a Boltzmann distribution, and the temperatures T_e, T_i, T_n are computed according to the analytical formulas $T(z) = T_\infty - (T_\infty - T_0)\exp\left(\frac{g}{RT_\infty}(z - z_0)\right)$ in good agreement with the experimental data. The constants in these formulas are taken as $T_{n\infty} = 800$ K, $T_{i\infty} = 950$ K, $T_{e\infty} = 2200$ K.

The first step is the change from the vector form of (2) to geographical coordinates in the assumption of a thin vertical layer (λ, φ, z), where z is the altitude counted off from the Earth's radius a. In the considered coordinates the expression for the divergence is $\operatorname{div}(n\vec{a}) = \dfrac{1}{a\cos\varphi}\left[\dfrac{\partial}{\partial\lambda}(na_x) + \dfrac{\partial}{\partial\varphi}(na_y\cos\varphi)\right] + \dfrac{\partial}{\partial z}(na_z)$, and the magnetic field components in the dipole approximations are $\vec{B} = (0, B\cos I, -B\sin I)$, where I is the magnetic inclination angle $I \approx \arctan(2\tan\varphi)$.

In the considered coordinates equation (2) is given by

$$\frac{\partial n}{\partial t} = DYZ(n) + DTr(n) + Tr(n) + [P - kn], \tag{3}$$

where the terms in the right hand side are

$$Tr(n) = \frac{1}{a\cos\varphi}\frac{\partial}{\partial\lambda}\left[n\frac{1}{B}(E_y \sin I + E_z \cos I)\right] + \frac{1}{a\cos\varphi}\frac{\partial}{\partial\varphi}\left[\left(u_z \sin I \cos I - u_y \cos^2 I - \right.\right.$$
$$\left.\left. -\frac{E_x}{B}\sin I\right) n\cos\varphi\right] + \frac{\partial}{\partial z}\left[\left(u_y \cos I \sin I - u_z \sin^2 I - \frac{E_x}{B}\cos I\right) n\right]; \tag{4}$$

$$DYZ(n) = \frac{1}{a\cos\varphi}\frac{\partial}{\partial\varphi}\left(D\cos\varphi\left[\frac{1}{a}\frac{\partial n}{\partial\varphi}\cos^2 I - \frac{\partial n}{\partial z}\cos I \sin I\right]\right) +$$
$$+\frac{\partial}{\partial z}\left(D\left[\frac{\partial n}{\partial z}\sin^2 I - \frac{1}{a}\frac{\partial n}{\partial\varphi}\cos I \sin I\right]\right); \tag{5}$$

$$DTr(n) = \frac{1}{a\cos\varphi}\frac{\partial}{\partial\varphi}\left[\left(\frac{1}{a}\frac{1}{T_p}\frac{\partial T_p}{\partial\varphi}\cos^2 I - \frac{1}{T_p}\frac{\partial T_p}{\partial z}\cos I \sin I - \right.\right.$$
$$\left.\left. -\frac{1}{H}\sin I \cos I\right) Dn\cos\varphi\right] + \frac{\partial}{\partial z}\left[\left(-\frac{1}{a}\frac{1}{T_p}\frac{\partial T_p}{\partial\varphi}\cos I \sin I + \frac{1}{T_p}\frac{\partial T_p}{\partial z}\sin^2 I + \frac{1}{H}\sin^2 I\right) Dn\right]. \tag{6}$$

A detailed physical validation of the model and derivation of the equations are beyond this work and will be given in a separate article.

2.2. Properties of the differential problem

1. The model is based on a mass conservation law taking place in absence of sources and sinks, which means that the problem has a corresponding conservation principle.
2. According to its physical sense the solution of the equation is non-negative (since the initial conditions are always chosen to be non-negative).
3. Near the lower boundary the diffusion component and transfer are negligible with respect to the photochemistry processes. In contrast, at the upper boundary the diffusion processes are dominating, and photochemistry plays no role. An important characteristic of the problem is a great variation of the parameters D, P, k, u in the considered region (by several orders of magnitude) [13]. The characteristic times of different physical processes vary considerably, and this causes a strong stiffness in the problem.

3. Solution method and numerical modeling

Consideration of the problem properties, listed in the previous section, provides the following requirements on the finite-difference schemes:

1. In order to reproduce the mass conservation law, only conservative finite difference schemes must be used.
2. The non-negativity of the solution determines a necessity to use monotone schemes (in the sense of Godunov's definition).
3. Since the overall goal is to build a coupled thermosphere-ionosphere INM RAS model [1], it creates a requirement to use big time steps in order to reduce the computational load.
4. Since the equation is stiff, it is necessary to use implicit time discretization schemes (taking into account the previous requirement).

3.1. Splitting method

The main method used in this work is a splitting method. Specificity of our approach lies in the fact that in order to simplify the numerical solving procedure of the 3D equation (3) splitting is used with respect to the physical processes and the geometric variables. For this, in equation (3) major physical processes (forming the concentrations and velocity fields), a vertical direction (connected with the gravitation field), and a meridional direction (connected with the magnetic field in the mentioned approximation) are chosen. In this work, only photochemistry processes and diffusion transfer are considered (the first two terms and the last one in equation (2)). The first splitting step describes plasma-chemical processes and diffusion in a projection on the z-axis:

$$\frac{\partial n}{\partial t} = \frac{\partial}{\partial z}\left[D\sin^2 I\left(\frac{\partial n}{\partial z}+\left(\frac{1}{T_p}\frac{\partial T_p}{\partial z}+\frac{1}{H}\right)n\right)-\frac{1}{a}D\sin I\cos I\left(\frac{\partial n}{\partial \varphi}+\frac{1}{T_p}\frac{\partial T_p}{\partial \varphi}n\right)\right]+[P-kn]. \quad (7)$$

At the second splitting step, an equation describing diffusion along the meridional direction is solved:

$$\frac{\partial n}{\partial t} = \frac{1}{a^2\cos\varphi}\frac{\partial}{\partial\varphi}\left[D\cos^2 I\frac{\partial n}{\partial\varphi}\cos\varphi\right] - \frac{1}{a\cos\varphi}\frac{\partial}{\partial\varphi}\left[D\sin I\cos I\frac{\partial n}{\partial z}\cos\varphi\right] +$$
$$+\frac{\partial}{\partial\varphi}\left[\left(\frac{1}{a}D\cos^2 I\frac{1}{T_p}\frac{\partial T_p}{\partial\varphi} - D\sin I\cos I\left(\frac{1}{T_p}\frac{\partial T_p}{\partial z}+\frac{1}{H}\right)\right)n\cos\varphi\right], \quad (8)$$

As a result of the two splitting steps, equation (3) is solved in 2D approximation without neutral and transversal transfer:

$$\frac{\partial n}{\partial t} = DYZ(n) + DTr(n) + [P - kn]. \quad (9)$$

The third splitting step governs 3D transfer, $Tr(n)$.

3.2. Implementation of the first splitting step

In this work, the results of numerical modeling of only the first splitting step are presented. This approximate formulation has a physical sense in the middle latitudes and near the poles without perturbations [13]. Solving the problem at the first splitting step is performed in several approximate settings. In the first setting, a 1D equation including plasma chemistry and diffusion along z is considered:

$$\frac{\partial n}{\partial t} = P - kn + \frac{\partial}{\partial z}\left[D\left(\frac{\partial n}{\partial z}+\left(\frac{1}{T_p}\frac{\partial T_p}{\partial z}+\frac{m_i g}{2kT_p}\right)n\right)\right] \quad (10)$$

with a Dirichlet boundary condition $n_{z=z_b} = \dfrac{P(z=z_b)}{k(z=z_b)}$ at the lower boundary $z = z_b$ and a constant flux condition $D\left(\dfrac{\partial n}{\partial z}+\left(\dfrac{1}{T_p}\dfrac{\partial T_p}{\partial z}+\dfrac{m_i g}{2kT_p}\right)n\right) = F$ at the upper boundary $z = z_t$ (in this work, for simplicity the constant F is approximately taken 0).

In the next approximate setting, the latitudinal dependency is taken into account by considering a projection on the z-axis (in this case the diffusion coefficient gets a multiplier $\sin^2 I$):

$$\frac{\partial n}{\partial t} = P - kn + \frac{\partial}{\partial z}\left[D\sin^2 I\left(\frac{\partial n}{\partial z}+\left(\frac{1}{T_p}\frac{\partial T_p}{\partial z}+\frac{m_i g}{2kT_p}\right)n\right)\right]. \quad (11)$$

The third approximate setting takes into account the latitudinal dependency, which leads to the 2D equation (7).

In this investigation, all external parameters of the model (neutral concentrations, temperatures, etc.) are defined with empirical formulas (see Appendixes in [13]).

3.3. Approximation method

The main approximation method in this work is a finite difference method. We now consider spatial approximations of each term in the approximation settings (7), (10), (11). At point $z = z_i$ for the diffusion component $\frac{\partial}{\partial z} D \frac{\partial n}{\partial z}$ the following standard conservative second order approximation is used:

$$\frac{\partial}{\partial z} D \frac{\partial n}{\partial z} \approx \frac{D_{i+1/2}(n_{i+1} - n_i)}{h^2} - \frac{D_{i-1/2}(n_i - n_{i-1})}{h^2}. \tag{12}$$

Let $u = D\left(\frac{1}{T_p}\frac{\partial T_p}{\partial z} + \frac{m_i g}{2kT_p}\right)$ denote an effective velocity. Then $\frac{\partial}{\partial z}\left(D\left(\frac{1}{T_p}\frac{\partial T_p}{\partial z} + \frac{m_i g}{2kT_p}\right)n\right)$ can be simply written as $\frac{\partial}{\partial z}(n \cdot u)$. For this component, central difference and directed difference schemes are used. The lower boundary condition is approximated exactly, and the upper boundary condition is approximated consistently with the whole finite difference scheme.

The schemes mentioned above are used in numerical implementation of the 1D equation (10). We have shown that these schemes are conservative and the directed difference scheme is monotone in the hyperbolic component. The basic idea of the conservation proof is considering a finite-dimensional analogue of the integral of the electron concentration over the whole area, calculated with corresponding quadratures and showing that its time derivative is zero. For a proof of the scheme monotonicity it can be shown that the matrix of the finite-dimensional operator is an M-matrix, whence it follows that its inverse matrix has non-negative elements. Therefore, the solution holds its sign passing to the next time layer.

For equation (11) with the projection on the magnetic field lines (by adding a factor $\sin^2 I$), the same finite difference schemes are used with latitude φ added as an external parameter.

For the equation with a projection and a mixed derivative (7), the main idea for building finite difference schemes is based on representing the mixed derivative in the following form:

$$\frac{\partial}{\partial z}\frac{\partial n}{\partial \varphi} = \frac{\partial}{\partial z}\left(n \frac{1}{n}\frac{\partial n}{\partial \varphi}\right) = \frac{\partial}{\partial z}\left(n \frac{\partial \ln n}{\partial \varphi}\right). \tag{13}$$

We denote

$$u_\varphi = \frac{\partial \ln n}{\partial \varphi} = \frac{1}{n}\frac{\partial n}{\partial \varphi}. \tag{14}$$

For equation (7) u_φ is an additive to the effective velocity u corresponding to the mixed derivative. Formally, the linear mixed derivative is represented as a divergence transfer with an effective velocity depending on the solution. This velocity, u_φ, can be calculated explicitly with the following finite difference schemes:

1. The first method is approximating the derivative of $\ln n$ by φ with central or directed difference in φ. For the central difference scheme this approximation is

$$u_\varphi \approx \frac{\ln n_i^j(\varphi + \Delta\varphi) - \ln n_i^j(\varphi - \Delta\varphi)}{2\Delta\varphi}; \tag{15}$$

2. The second method is approximating this derivative without the ln function:

$$u_\varphi \approx \frac{2}{n^j(\varphi + \Delta\varphi) + n^j(\varphi - \Delta\varphi)} \cdot \frac{n^j(\varphi + \Delta\varphi) - n^j(\varphi - \Delta\varphi)}{2\Delta\varphi}. \tag{16}$$

To approximate the mixed derivative term $\frac{\partial}{\partial z}(u_\varphi n)$ in z, two conservative schemes of the form $\frac{1}{h}(\Phi_{i+1/2} - \Phi_{i-1/2})$ are explored:

1. Central difference approximation:

$$\Phi_{i+1/2} = \frac{u_{\varphi(i+1)}n_{i+1} + u_{\varphi(i)}n_i}{2}; \tag{17}$$

2. Directed difference approximation with a flux calculated with

$$\Phi_{i+1/2} = \frac{u_{\varphi(i+1/2)} + |u_{\varphi(i+1/2)}|}{2h}n_i + \frac{u_{\varphi(i+1/2)} - |u_{\varphi(i+1/2)}|}{2h}n_{i+1}. \tag{18}$$

In numerical modeling of all considered approximation settings implicit time discretization schemes are used, according to the above-mentioned requirement on the computational efficiency of the solving methods. In this case the time step corresponds to the time step used in the INM RAS thermosphere model (1-5 minutes). The mixed derivative term is approximated with an explicit-implicit scheme: the effective additive u_φ that contains a latitudinal derivative is calculated with an explicit scheme, and the full term is approximated with an implicit one.

In the above-considered assumptions for computing the solution at the first splitting step, the following boundary conditions are used: at the lower z-boundary we have a Dirichlet boundary condition $n(z_b) = \dfrac{P(z_b)}{k(z_b)}$, and at the upper boundary, as before, the boundary condition is a constant flux, but this time with the addition of u_φ.

4. Results

We now proceed to the results of numerical experiments for the first version of the model in different approximation settings and their comparison. A detailed investigation is provided for convergence of the numerical solution to a stationary distribution in case of fixed daytime photoionization P. The spatial characteristics of the stationary solution of the finite difference problem within implementation of the finite difference schemes for the three main approximation settings $(10, 11, 7)$ are also considered.

The numerical experiments in the approximation settings $(10, 11)$ were performed with a space step $h = 5$ km and a time step $\tau = 3$ min. These values fit the properties of the INM RAS thermosphere model [1]. As for equations $(11, 7)$, a latitudinal step of 1° was used.

The numerical experiments in settings without the mixed derivative $(10, 11)$ have shown convergence to a stationary distribution that is close to the observed unperturbed electron concentration distribution in the terrestrial F layer in the middle latitudes [13]. The characteristic stabilization time has an order of $4 - 5$ hours. Using various finite difference schemes for the hyperbolic part did not lead to a significant change in the results.

A comparison of the results of the numerical experiments on the reconstitution of a daytime stationary vertical distribution for the main approximation settings $(10, 11, 7)$ in different latitudes is given in Figure 1. It can be easily seen that the largest differences (as it follows immediately from the equations) are observed near the equator, whereas the closest solution to the characteristic distribution in the terrestrial ionosphere in this case is obtained from equation (10).

A distinctive feature of equations (11) and (7) (taking into account the inconsistency of the vertical and magnetic field directions) is a singular non-physical photochemistry solution at the equator: $n = \dfrac{P}{k}$. As a consequence, near the equator the solution has a significant latitudinal gradient. Numerical experiments on solving (7) with the mixed derivative have shown that the central difference scheme $(15, 16)$ in φ gives an intense false two-step oscillation spreading from the equator with a rising amplitude. The implementation of the directed difference formula in φ, on the contrary, did not cause any significant problems. A remarkable fact is that there is

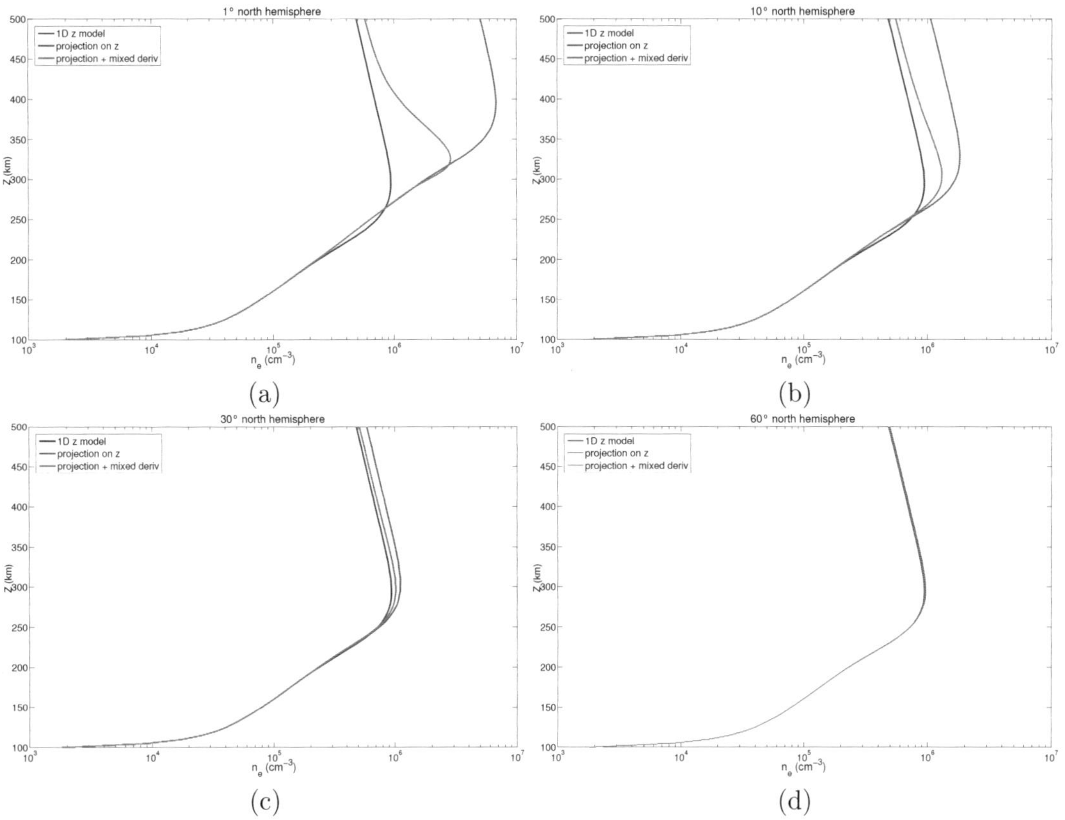

Figure 1. Vertical distributions of the electron concentration (cm^{-3}, in log scale) subsequent to the results from three approximation settings (blue — (10), green — (11), red — (7)) at different latitudes: (a) $-1°$, (b) $-10°$, (c) $-30°$, (d) $-60°$ north latitude (for the south latitude the solution is the same according to the symmetry of the equation).

no observable difference between the approximations of u_φ with the logarithm and the simple directed difference (15, 16).

Figure 2 illustrates the results of calculating the height-latitudinal distributions in approximation settings without the mixed derivative term (11) and with this term (14) after the calculation for a day. It is easy to see that adding the mixed derivative component can have an impact not only near the equator, but also in subtropical zones up to the middle latitudes.

Despite limitations, the above-considered first version of the model gives a possibility to investigate the primary physical problem of sensitivity to parameters of the equation which are external for the ionosphere. Figure 3 shows the vertical electron concentration profiles for the 1D equation (10) with variation of the following parameters: $n_O(z_0)$ in Boltzmann distribution, temperatures $T_{n\infty}$ and $T_{e\infty}$ at the upper boundary of the considered area, and photoionization P. It was demonstrated that the profile has the highest sensitivity to the neutral temperature, the ionization level, and the atomic oxygen concentration. The other parameters have a slight impact on the solution. It is important to notice that the specific variability structure of n by virtue of the neutral temperature (mainly variation of the altitude of the F region maximum) is qualitatively different from the structure corresponding to the photoionization (variation of the maximum value of the F region). This result might be of importance for understanding the mechanisms of ionosphere variability.

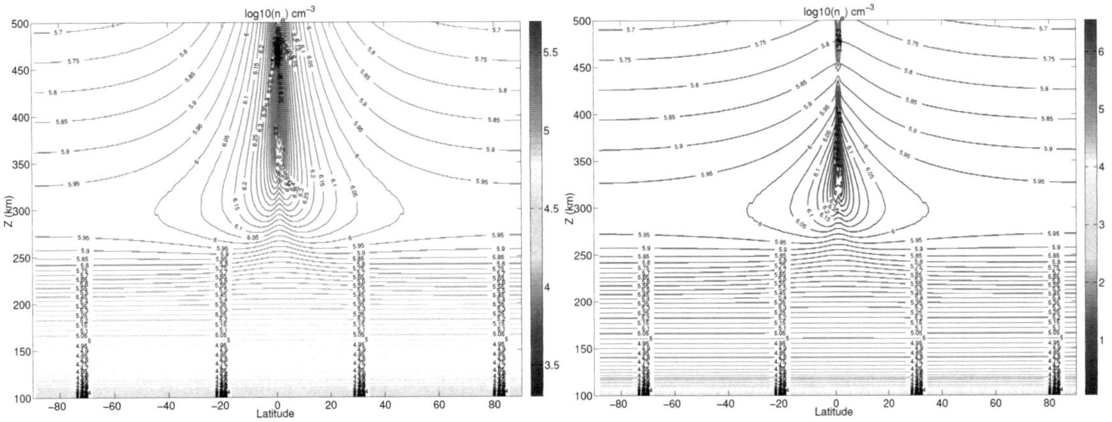

(a) (b)

Figure 2. Height-latitudinal distributions of electron concentration (decimal logarithm, cm^{-3}) according to the results of the created ionosphere model in the approximation setting with vertical direction projection without the mixed derivative ((11) (a)) and with it ((7) (b)). Isolines are drawn with a step of 0.05.

5. Conclusions

In conclusion, we briefly outline the basic results.

A complete 3D model of the terrestrial ionosphere's F layer was stated and formulated in spherical layer coordinates. The model includes the main physical processes forming the ionosphere. An original method was suggested for a stage-by-stage realization of the model with splitting with respect to the physical processes and the geometric variables. The first splitting step for the vertical direction was implemented in three approximation settings (considering only the vertical diffusion, taking into account a projection on the magnetic field lines without or with a mixed derivative term). Several finite difference schemes were implemented: the diffusion part was approximated with a central difference scheme and for the hyperbolic transfer central and directed difference approximations were employed. The conservation and monotonicity were shown for a directed difference scheme for the first setting.

An essentially novel technique in the context of numerical methods was used in solving the equation with the mixed derivative. Several original ways of approximating this derivative were suggested. It was also shown that the central difference approximation along one of the directions can give significant numerical errors and is not recommended in case of expecting strong gradients of the solution.

Generally, an analysis of the obtained results has shown the applicability of the discussed approximation setting to modeling of the ionosphere's F layer in an unperturbed condition in the middle latitudes and the necessity to consider meridional diffusion in a region near the equator where the magnetic field is approximately horizontal. This consideration will be the first stage of future work.

Through the use of the dynamical model in the middle latitudes the sensitivities of the stationary electron concentration vertical distribution to external equation parameters were investigated. The most remarkable result (from a physical standpoint) is that the greatest relative influence on the concentration distribution has been exerted by the neutral temperature, atomic oxygen concentration, and ionization level. Therewith the spatial variability structures determined by the mentioned variations are qualitatively different.

According to the successful implementation of the first splitting step in the above-suggested scheme, a general strategy of solving the problems stated in the introduction includes stepwise

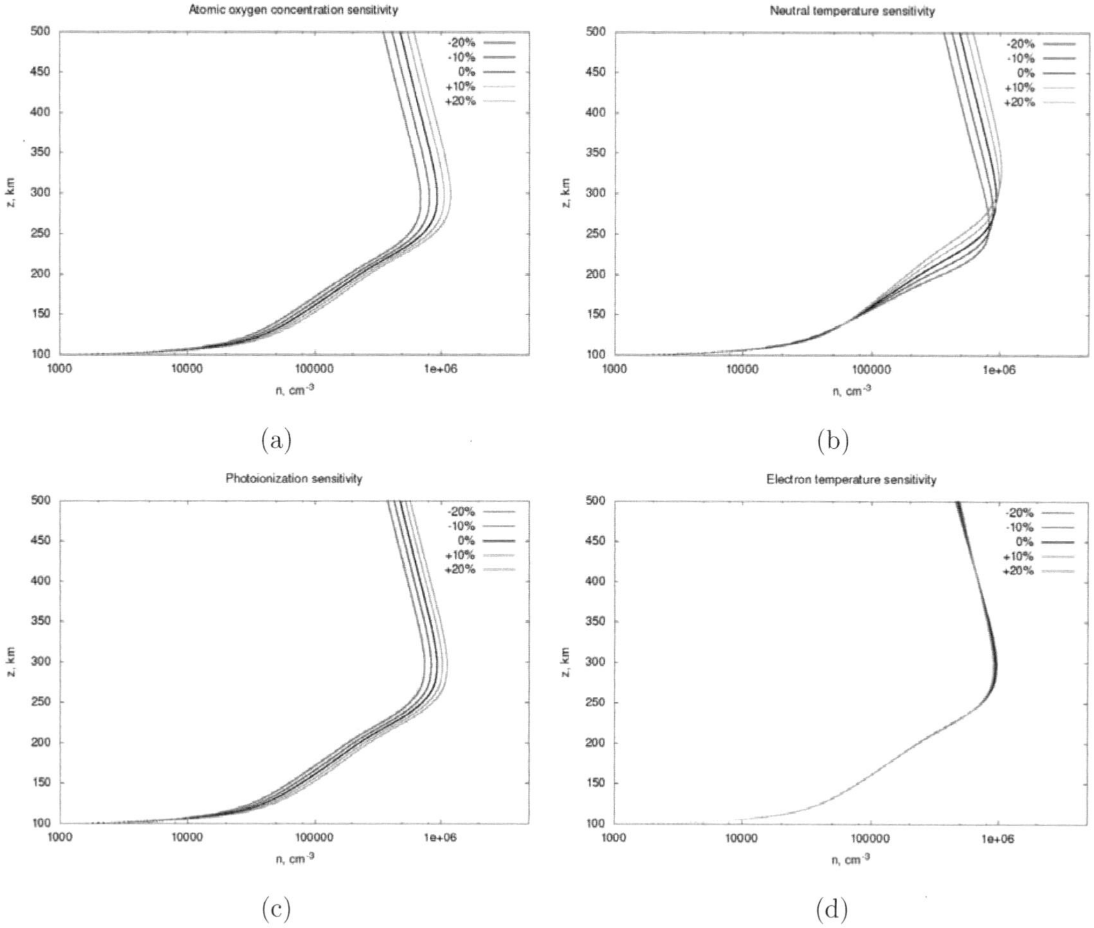

Figure 3. Electron concentration vertical distribution (cm^{-3}, in log scale) obtained by variation of the parameters external for the ionosphere model (10) by 10% and 20% on either side: (a) atomic oxygen concentration, (b) neutral temperature, (c) photoionization P, (d) electron temperature.

solving of the complete 3D equation based on the above-discussed algorithms and an integration of the model into a developed thermosphere's model.

Acknowledgments
The work was supported by Russian Science Foundation (project no. 17-17-01305).

6. References
[1] Kulyamin D V and Dymnikov V P 2013 A three-dimensional model of general thermospheric circulation *Russian J. Numer. Anal. Math. Modelling* **28**(4) 353-80
[2] Kulyamin D V and Dymnikov V P 2015 Modelling of the lower ionosphere climate. *Izv. - Atm. Ocean. Phys.* **51**(3) 272-91
[3] Bilitza D, McKinnell L A, Reinisch B and Fuller-Rowell T 2011 The International Reference Ionosphere (IRI) today and in the future *J. Geodesy*, **85** 909-20
[4] Leshchinskaya T Y, Mikhailov V V 2016 SIMP-1 model: mapping of fo F2 monthly medians over the Northern *Geomagnetism and Aeronomy* **56**(6) 733-41

[5] GOST 25645.146-89 *The Earth's ionosphere. Model of global distribution of concentration, temperature, and effective collision rate of electrons* (Moscow: Izdatel'stvo standartov) 1990
[6] Richmond A D, Ridley E C and Roble R G 1992 A Thermosphere/Ionosphere General Circulation Model with coupled electrodynamics *Geophys. Res. Lett.* **19** 601-4
[7] Roble R G, Ridley E C, Richmond A D and Dickinson R E 1988 A coupled thermosphere/ionosphere general circulation model *Geophys. Res. Lett.* **15** 1325-8
[8] Bailey G J, Balan N and Su Y Z 1997 The Sheffield university plasmasphere ionosphere model: a review *Journal of Atmospheric and Solar-Terrestrial physics* **59**(13) 1541-52
[9] Namgaladze A A, Martynenko O V, Volkov M A, Namgaladze A N and Yurik R Yu 1998 High-latitude version of the global numerical model of the Earth's upper atmosphere *Proceedings of the MSTU.* **1**(2) 23-84
[10] Surotkin V A, Klimenko V V and Namgaladze A A 1979 A numerical model of the equatorial ionosphere *Investigation of the ionospheric dynamics* 58-68
[11] Schunk R W, Scherliess L, Sojka J J, Thompson D C, Anderson D N, Codrescu M, Minter C, Fuller-Rowell T J, Heelis R A, Hairston M and Howe B 2004 Global assimilation of ionospheric measurements (GAIM) *Radio Sci.* **39** RS1S02
[12] Khattatov B, Murphy M, Gnedin M, Sheffel J, Adams J, Cruickshank B, Yudin V, Fuller-Rowell T and Retterer J 2005 Ionospheric nowcasting via assimilation of GPS measurements of ionospheric electron content in a global physics-based time-dependent model *Q. J. R. Meteorol. Soc.* **131** 3543-59
[13] Schunk R W and Nagy A F 2009 *Ionospheres Physics, Plasma Physics, and Chemistry* (New York: Cambridge University Press)

Architecture of a spatial data service system for statistical analysis and visualization of regional climate changes

A G Titov[1,2], I G Okladnikov[1,2] and E P Gordov[1,2]

[1] Institute of Monitoring of Climatic and Ecological Systems SB RAS, Tomsk, Russia
[2] V.E. Zuev Institute of Atmospheric Optics SB RAS, Tomsk, Russia

titov@scert.ru

Abstract. The use of large geospatial datasets in climate change studies requires the development of a set of Spatial Data Infrastructure (SDI) elements, including geoprocessing and cartographical visualization web services. This paper presents the architecture of a geospatial OGC web service system as an integral part of a virtual research environment (VRE) general architecture for statistical processing and visualization of meteorological and climatic data. The architecture is a set of interconnected standalone SDI nodes with corresponding data storage systems. Each node runs a specialized software, such as a geoportal, cartographical web services (WMS/WFS), a metadata catalog, and a MySQL database of technical metadata describing geospatial datasets available for the node. It also contains geospatial data processing services (WPS) based on a modular computing backend realizing statistical processing functionality and, thus, providing analysis of large datasets with the results of visualization and export into files of standard formats (XML, binary, etc.). Some cartographical web services have been developed in a system's prototype to provide capabilities to work with raster and vector geospatial data based on OGC web services. The distributed architecture presented allows easy addition of new nodes, computing and data storage systems, and provides a solid computational infrastructure for regional climate change studies based on modern Web and GIS technologies.

1. Introduction

Dramatically fast global climate changes have been observed during the last 50 years. They set conditions for rapid development of environmental monitoring and modeling technologies. New methods of processing of environmental geospatial datasets have been developed, and most of them are based on modern data science methods. Among them are statistical methods of extracting extreme climate events from meteorological time series presented in the form of geospatial datasets, as well as assessment of their spatio-temporal dynamics. The use of large geospatial datasets in climate change studies requires performing a number of procedures, such as dataset collection, storage, search, exchange, transfer, processing, visualization, and analysis. To achieve this, a set of Spatial Data Infrastructure (SDI, [1]) elements should be created and published, particularly including geophysical data processing (geoprocessing) web services and services of cartographical visualization. It is generally accepted that the development of client applications as integrated elements of such an infrastructure should be based on the use of modern Web and GIS technologies [2, 3, 4, 5]. According to the general requirements of the INSPIRE Directive on geospatial data visualization [6], it is necessary to provide such features as data overview, image navigation, scrolling, zooming and graphical overlay, as well as displaying map legends and related meta information. Also, the OGC specifications define that the basic functions to be performed by geoprocessing services are as follows: geospatial data visualization (WMS), provision in vector (WFS) and raster (WCS) formats, as well as geospatial analytical processing (WPS).

Content from this work may be used under the terms of the Creative Commons Attribution 3.0 licence. Any further distribution of this work must maintain attribution to the author(s) and the title of the work, journal citation and DOI.
Published under licence by IOP Publishing Ltd

IOP Conf. Series: Earth and Environmental Science **96** (2017) 012012 doi:10.1088/1755-1315/96/1/012012

At present there are a number of information systems and services that provide some parts of the functionality required. An advanced example is a fully integrated system created at the University of New Hampshire for online analysis of heterogeneous climate, hydrology, and remote sensing data (RIMS). It is a successful attempt to implement a multi-functional GIS as a web application using MapServer (http://mapserver.org/) software. RIMS have been widely used in various research projects [7, 8] including NEESPI (Northern Eurasian Earth Science Partnership Initiative, http://NEESPI.sr.unh.edu/). A major feature of RIMS is that it deals directly with geospatial datasets instead of corresponding cartographical images, and allows access to the data displayed at each pixel of the map, as well as search and access to data aggregated according to some space and/or time criteria. Its main defect that narrows the system's possible use is that it does not meet the SDI/OGC standards.

This paper presents the architecture of a geospatial OGC web service system as an integral part of a virtual research environment (VRE) general architecture for statistical processing and visualization of meteorological and climatic data aimed at providing decision makers and researchers with accurate and reliable climate change characteristics. The architecture can be considered as a result of an elaborated synergy approach based on RIMS technological solutions and techniques developed by the authors.

2. General architecture of the VRE

The architecture (Figure 1) represents a set of interconnected standalone SDI nodes (Virtual Research

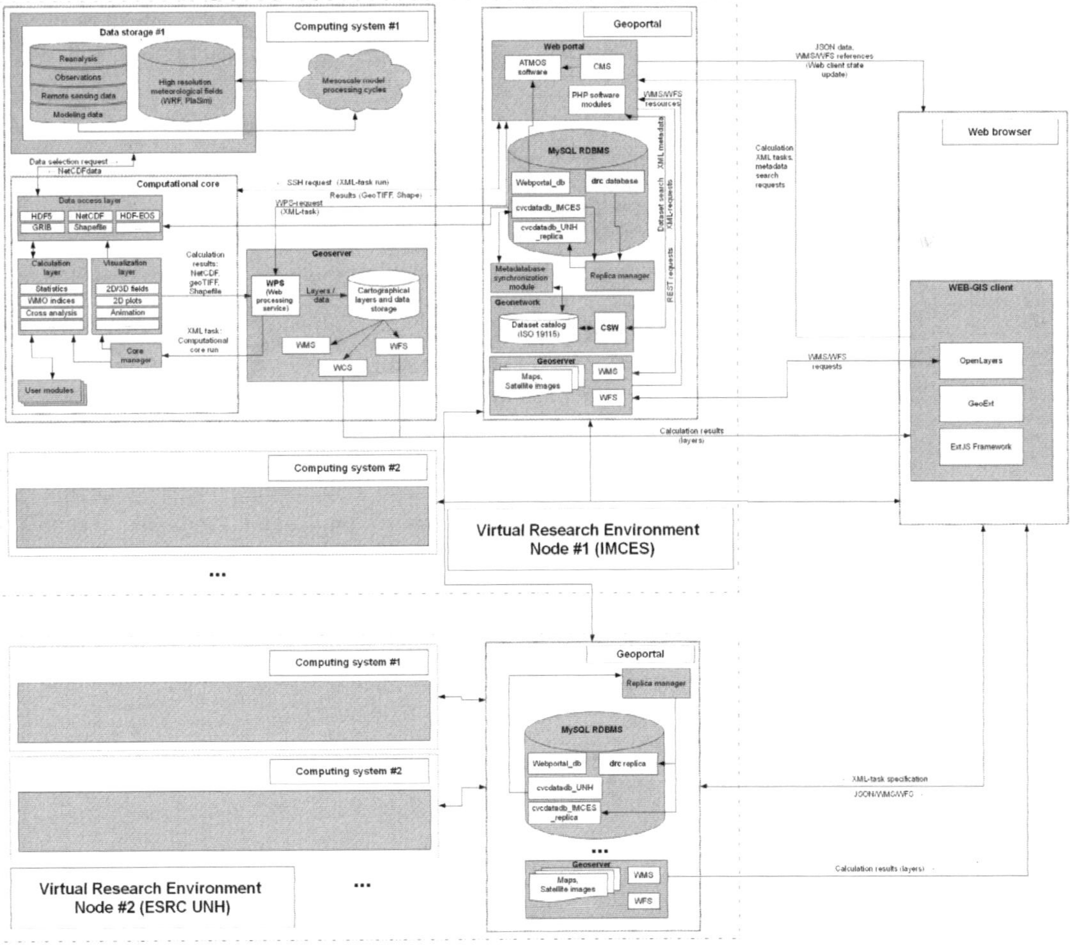

Figure 1. General architecture outline.

Environment Node #1 and Node #2 are displayed as examples) with corresponding computing systems (servers) and data storage systems containing geospatial datasets basically in the NetCDF format, including numerical modeling and reanalysis data, as well as meteorological observation data grouped according to thematic characteristics. Each node runs a specialized software, such as an SDI geoportal, cartographical web services (WMS/WFS) based on Geoserver (http://geoserver.org/), a MySQL database of technical metadata describing the geospatial datasets available for the node, and geospatial data processing services (WPS) based on a modular computational backend realizing statistical processing functionality and thus providing access to, as well as complex processing and analysis of large datasets, results of visualization and export into files of the required formats. Within the general architecture framework, the nodes' MySQL metadata databases and their interconnections are of fundamental importance. The metadata database of the central Node #1 (drc_database) contains descriptions of all the SDI nodes available in the environment as a whole, while each of them, in turn, is connected to a set of computing servers (Computing system #1, ...), data storages and, finally, the available datasets. The SDI node descriptions consist of such fields as name, title, URL, and node's geoportal MySQL database connection parameters. All other secondary nodes contain a central metadatabase replica that allows geoportal applications to check the information about datasets available for processing on other nodes, as well as computing modules for their processing. The central metadata database can be administered by a senior system administrator, while the nodes' administrators have "read-only" access rights for it and its replicas. Due to this approach, the VRE users are provided with information about all the geographically distributed datasets globally accessible, regardless of the availability of direct network connections to the remote databases at any time. Obviously, a standalone node is characterized by the presence of: at least one high-performance computing system with a set of data storage attached; an SDI geoportal; and a computational backend.

The architecture at a single VRE node level may be represented by three basic tiers:

1. Tier of data and calculation procedures providing analysis and corresponding cartographical and visualization services.
2. Tier of middleware representing the geoportal including metadata database and server-side web applications.
3. Tier of client-side web-applications providing the GIS functionality (Web-GIS client, etc.) available to the end user.

2.1. Tier of data and calculation procedures

The tier of data and calculation procedures provides analytical processing of geospatial datasets, as well as cartographical and visualization services. The datasets are located on data storage systems either in the form of collections of netCDF files or PostGIS databases.

2.1.1. Modular computational backend. Within the framework of the data tier, the key software component is a specialized computational backend representing a standalone software toolset. The computational backend of the VRE is based on the GNU Data Language (GDL, http://gnudatalanguage.sourceforge.net/) and Python (http://python.org) providing integral geospatial data mathematical processing and visualization functionality, as well as API to work with NetCDF/HDF, Esri Shapefile data files, and PostGIS databases. The computational backend contains the following key components:

1. Data Input / Output component
2. Geospatial data processing component
3. Graphical visualization component

Depending on the result type required, the backend generates the following file formats: GeoTIFF, ESRI Shapefile, Encapsulated PostScript, CSV, XML, NetCDF, and float GeoTIFF. The computational backend manager controls the general application workflow, thus providing data processing modules calls and their interaction. Computational pipelines are prepared with a special XML file describing a computational task generated by a Web GIS client according to the actions of the end user.

2.1.2. OGC web services. The geospatial data service system is supposed to provide launching of computational procedures and visualization of their results. The cartographical web services to be realized as an integral part of the system implement the differentiation of the user's access rights and provide functional capabilities to work with raster (GeoTIFF) and vector (Shapefile, PostGIS) geospatial data using OGC WMS, WFS, and WPS based on the Geoserver and GeoWebCache software (http://geoserver.org). In general, the cartographical Web processing service provides a standard HTTP interface for remote configuring and launching the data processing modules and presenting the results in binary and graphical formats according to the end user's instructions. Additionally, the task of transforming graphical calculation results along with SLD styles into corresponding Geoserver cartographical layers and styles, as well as their publishing as WMS and WFS resources, should be solved. As successful criteria of the resource publishing, the following might be used:

1. The presence of the WMS/WFS resource descriptions of calculation results in OGC GetCapabilities replies
    a. Obtaining of graphical legends in reply to the GetLegendGraphic WMS request
    b. Correct visualization of the cartographical layer with the required SLD style applied when using any client GIS application (OpenLayers, uDIG, QGIS)
2. Classical differentiation of the Geoserver user access rights for view, creating and modifying of WMS and WFS resources must be provided with all the users, except for the administrator group members able to work with their own cartographical resources.

Thus, the services developed can be used either while working in the standard GIS environment (uDIG, QGIS, etc.) or in specialized web applications deployed within the geoportal framework.

2.2. Tier of middleware

According to the Spatial Data Infrastructure concept, the geoportal serves as a gateway to the geospatial data and derived products, and provides functional capabilities, such as search for geoinformation resources using catalogs, data access, and sampling, as well as management of the geoprocessing services and cartographical processing and visualization applications [9]. The virtual Research Environment geoportal represents a middleware consisting of:

1. A metadata database describing the available geospatial datasets.
2. A metadata catalog based on GeoNetwork to describe the geoinformation resources used in climate research using an ISO 19115 standard.
3. A web portal implementing: web applications as reusable PHP modules and interconnections with the OGC web services, target computing systems, and a metadata database.
4. A central Geoserver repository containing basic cartographical layers, such as state and geographic boundaries, land cover and climate zone maps, river basins, etc.

2.2.1. Metadata database. At a single VRE node, its metadata database [10] presents multilingual descriptions of climatic geospatial datasets, with processing routines in the form of dedicated software modules. Thus, there are two major parts of the MDDB:

1. One part contains information about all datasets available for analysis. It is used to locate data files and to provide metadata on computational backend request. This part of the MDDB describes the geospatial datasets available for the node: their spatio-temporal characteristics, lists of meteorological parameters and physical locations, which provides fast location of data files on storage systems by the computational backend.

2. The other part contains descriptions of processing routines represented as various pipelined call sequences of dedicated computational backend modules and their run options. Since some data analysis routines are designed to process only specific meteorological parameters, the connections between the computing modules and data arrays are set in the MDDB. It also contains descriptions of the corresponding GUI elements required to provide the module calculation.

In Figure 1 the metadata database for a particular node of the VRE has such name like cvcdatadb_<# of the VRE node> (cvcdatadb_1 for the VRE node # 1, etc.), and except for the relations describing datasets and the corresponding routines, it has the following additional tables:

– computing_system_type: types of the available computing systems (local, remote, OGC WPS);
– computing_system: list of descriptions of the computing systems available for the VRE node.

All the other VRE nodes contain replicas of the considered metadatabase named cvcdatadb_<# of the VRE node>_replica, which provides the VRE users with the information about all geographically distributed datasets available within SDI regardless of the presence of Internet connection. Actualization of the replica is performed by the geoportal "Replica manager" component. The node's metadatabase can be administered by the local system administrator; the administrators of other nodes have only read access to it.

2.2.2. Web portal. A web portal complying with the general concepts and standards of development of an open source software providing OGC cartographical web services serves as a connection point between the SDI elements. It implements basic functionality, such as user's authentication, RDBMS connectivity, HTML templates usage, multilingual support, a content management system, and some other features [11]. The core part of the web portal was developed aiming at computational solution of scientific tasks in meteorology and climatology. Its main feature provides a unified API for web applications.

2.3. Tier of client-side web applications providing GIS functionality

The geoportal web applications comply with the conventional Boundless / OpenGeo architecture [12]. The architecture consists of three basic tiers (data, application server, and graphical user interface) and implies the use of the following open source software: PostGIS, Geoserver, OpenLayers, and GeoExt/ExtJS.

2.3.1. Web GIS client. The Web GIS client, as the main software instrument of the VRE desktop, consists of two basic software parts:

1. The server-side part as a set of web portal PHP applications implementing an open API for browser-based client software part and realizing the functionality of interaction with:
   a. GDL / Python backend in the form of a standalone computing core;
   b. WMS/WFS/WPS cartographical services based on the Geoserver software.

   This part has a secondary meaning, since it does not contain an application business logic, while providing a limited set of GUI manipulation procedures via the standard HTTP interface.
2. The front-end part representing web mapping client developed according to a "single page application" technology based on JavaScript OpenLayers (http://openlayers.org/) and GeoExt (http://geoext.org/) libraries and providing a user-friendly interface similar to one of the well-known desktop applications, such as uDIG, QGIS, and so on.

There are two main server-side web applications:

1. Manager of cartographical layers to administer Geoserver raster and vector layers via the standard REST (Representational State Transfer) interface. It also includes a processing unit for handling the computational task obtained from the user in XML format and the computational backend launching and tracking module.

2. Metadata manager working with geospatial dataset formalized descriptions developed to interact with the MySQL metadatabase and providing a presentation of the NetCDF metadata in JSON format optimized for processing by the browser based GIS client. The technique applied allows the client side structured data caching and updating of the graphical user interface elements.

The cartographical front-end web application (web mapping client) for work with the archive of geospatial NetCDF datasets contains 3 basic tiers [13]:

1. Tier of NetCDF metadata retrieved from MySQL database and represented in JSON format
    a. Middleware tier of JavaScript objects implementing methods to work with:
    b. JSON metadata
    c. XML file containing configuration of selected processing routines, input and output formats (XML task)
2. OGC WMS/WFS cartographical services
3. Graphical user interface tier representing JavaScript objects realizing general application business logic

Web GIS client must comply with the general INSPIRE requirements on geospatial data visualization and provide computational processing services launching to support the solving of tasks in environmental monitoring, as well as presenting the calculation results in the form of WMS/WFS cartographical layers in raster (PNG, JPG, GeoTIFF), vector (KML, GML, Shape), and binary (NetCDF) formats.

3. The prototype system

An element developed for the Virtual Research Environment is supported by two standalone nodes interconnected by telecommunication channels, which form a hardware complex providing distributed access, analytical processing, and visualization of large geospatial datasets using cloud computing techniques. An access to the VRE geospatial data services is provided via the standard web browser from any workstation with an Internet connection.

Within the framework of Web GIS client OGC WMS/WFS services the support is realized as an integrated component. Its goal is the interaction with the system end user during the digital map creation process, Geoserver connectivity, and accurate presentation of the calculation results using modern Web and GIS technologies. The component provides such basic elements as:

1. Window displaying lists of the available raster and vector layers accessible via WMS/WFS, as well as their preview
2. Module handling requests adding new layers to the map
3. Layer legend display panel
4. Module to work with WFS layers including element selection and attribute table retrieval
5. Window displaying results of the user GetFeatureInfo request (geographic position, parameter values for the point selected)

A general algorithm implemented by the component supporting OGC web services is presented below:

a) While creating a thematic map using Web GIS client the user opens a dialog window listing the WMS/WFS layers available
b) The user selects the layers to be added to the map
c) An HTTP request is formed and sent to the corresponding cartographical web service

d) According to the HTTP reply obtained, the web interface graphical components are modified and selected layers are displayed on the map in the correct order
e) The user analyzes the map and attribute tables of the point of interest
f) The user gets either a raster (GeoTIFF) or vector (GML, JSON) presentation of the selected cartographical layer and corresponding source data in NetCDF format.

The Web GIS client of the prototype was developed using DHTML, PHP, and JavaScript languages. The JavaScript library implementing Web GIS client interface components is based on OpenLayers, GeoExt and ExtJS libraries, and OGC services provided by Geoserver. Graphical interface performs two main functions: providing functional capabilities for editing calculation task XML file, and a visual presentation of the cartographical information for the end user. The basic elements of the graphical user interface are (Figure 2):

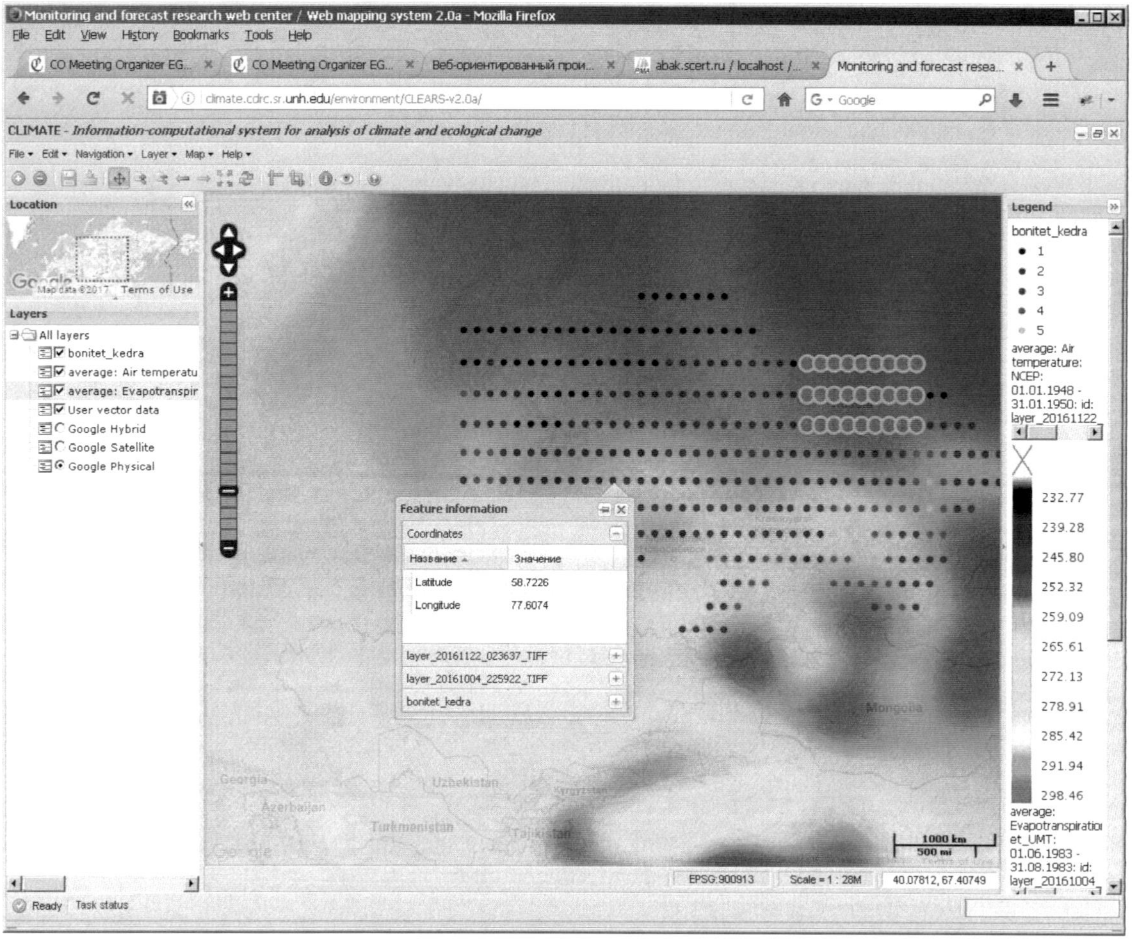

Figure 2. Web GIS client interface.

- Panel displaying user cartographical layers on the map, Google maps being used as the base layer by default, but there is a possibility to set an arbitrary base layer including a new one created by the user
- Layer tree allowing to toggle layer display
- Layer legends display panel

- Map information panel (scaling, cartographical projections, cursor geographical coordinates)
- Application general status panel
- Overview map panel
- General application menu
- Toolbar (adding/removing layer, saving NetCDF data, panning, map refresh, obtaining of information related to given geographical point, etc.)
- Application context menu
- Wizard creating cartographical layers (Figure 3) based on the results of computational processing of geospatial datasets available to the system

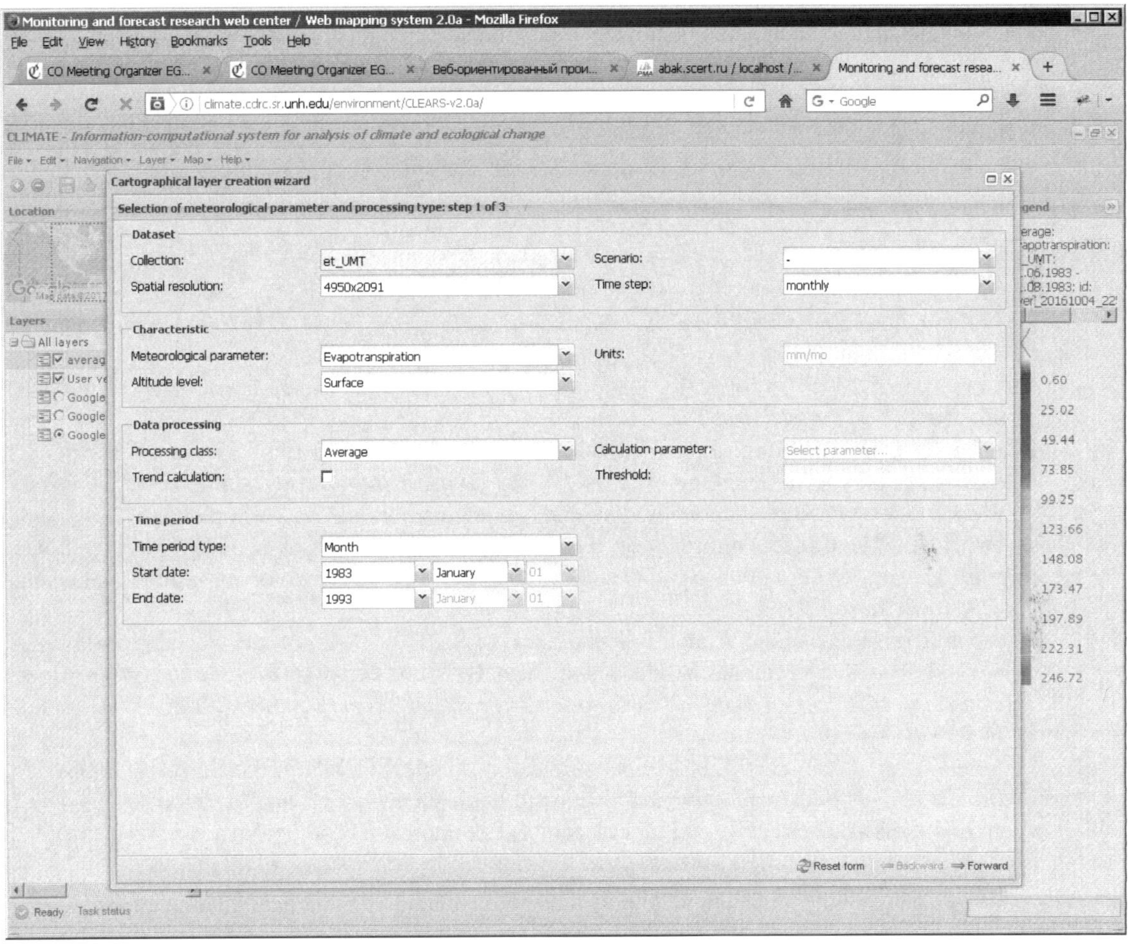

Figure 3. Layer creation wizard.

A cartographical layer creation wizard was developed based on the following criteria:

1. Efficiency of filling out the graphical user interface interactive forms;
2. Optimization of the process of creating and editing of an XML file describing the selected calculation configuration.

The toolbar, application and context menus contain mouse and keyboard event handlers which uniquely determine the Web-GIS behavior depending on the user's actions with the execution context applied. It should be noted that the geospatial data cartographical services based on the Geoserver software can be used in the Web-GIS client considered, as well as in the standard desktop GIS applications.

4. Conclusions

The VRE distributed architecture presented allows easy addition of new nodes, computing and data storage systems. It provides a solid computational infrastructure for regional climate change studies based on modern Web and GIS technologies. The metadata database has improved the system functional capabilities in terms of extending the geospatial dataset archives and the statistical processing routines, as well as providing computational resources as web services. The front-end developed complies with the general GIS requirements, while the computing backend technical realization allows a unified geospatial dataset processing regardless of the client application type. The working prototype presented has been shown to be effective in the process of solving real climate change research problems and disseminating the investigation results in cartographical form. It has been developed for decision makers and specialists working in affiliated sciences, and provides accurate environmental characteristics required in the studies of economic and social consequences of global climate change at the regional level.

Acknowledgments
The authors thank the Russian Science Foundation for the support of this work under grant # 16-19-10257.

References

[1] Steiniger S and Hunter A J S 2012 Free and open source GIS software for building a spatial data infrastructure *Geospatial Free and Open Source Software in the 21st Century* ed E Bocher and M Neteler (Heidelberg: LNGC, Springer) pp 247–61
[2] Yakubailik O E 2007 Geoformation geoportal *J. Computational Technologies (Special Issue 3)* **12** 116–25
[3] Dragicevic S, Balram S and Lewis J 2000 The role of Web GIS tools in the environmental modeling and decision-making process *4th Int. Conf. on Integrating GIS and Environmental Modeling (GIS/EM4): Problems, Prospects and Research Needs* (September 2-8, 2000, Banff, Alberta, Canada) http://www.srcosmos.gr/srcosmos/showpub.aspx?aa=5823
[4] Frans J M van der Wel 2005 Spatial data infrastructure for meteorological and climatic data *Meteorol. Appl.* **12** 7–8
[5] Vatsavai R R, Thomas E B, Wilson B T and Shekhar S 2000 A Web-based browsing and spatial analysis system for regional natural resource analysis and mapping *Proc. of the 8th ACM Int. symp. on Advances in geographic information systems* (Noveber 6-11, 2000, Washington, D.C., US) 95–101
[6] Janssen K 2010 The Availability of Spatial and Environmental Data in the European Union: At the Crossroads Between Public and Economic Interests *Kluwer Law International*, p 617
[7] Welp L R, Randerson J T, Finlay J C, Davydov S P, Zimova G M, Davydova A I and Zimov S A 2005 A high-resolution time series of oxygen isotopes from the Kolyma River: Implications for the seasonal dynamics of discharge and basin-scale water use *Geophys. Res. Lett.* **32** L14401 doi: 10.1029/2005GL022857.
[8] Shiklomanov A I 2012 Regional Integrated Monitoring System for the Pan-Arctic (RIMS) *Proc.of the Northern Eurasian Earth Science Partnership Initiative (NEESPI) Regional Science Team Meeting devoted to the High Latitudes* (June 2-6, 2012, Helsinki, Finland) 128–30
[9] Koshkarev A V 2008 Geoportal as a tool to control spatial data and services *Spatial data* **2** 6–14
[10] Okladnikov I G, Gordov E P and Titov A G 2016 Development of climate data storage and processing model *IOP Conf. Series: Earth Environ. Sci* **48** 012030
[11] Gordov E P Lykosov V N and Fazliev A Z 2006 Web portal on environmental sciences "ATMOS" *Adv. Geosci.* **8** 33-38

[12] Becirspahic L and Karabegovic A 2015 Web portals for visualizing and searching spatial data *Information and Communication Technology, Electronics and Microelectronics (MIPRO)* 305-11 doi: 10.1109/MIPRO.2015.7160284
[13] E Gordov, A Shiklomanov, I Okladnikov, A Prusevich and A Titov 2016 Development of Distributed Research Center for analysis of regional climatic and environmental changes *IOP Conf. Series: Earth Environ. Sci* **48** 012033

IOP Conf. Series: Earth and Environmental Science **96** (2017) 012013 doi:10.1088/1755-1315/96/1/012013

Cloud identification using actinometrical data

Sergey V Zuev[1], **Nikolay P Krasnenko**[1,2] **and Elena S Kartashova**[1]

[1] Institute of Monitoring of Climatic and Ecological Systems SB RAS, Tomsk, Russia
[2] Tomsk State University of Control Systems and Radioelectronics, Tomsk, Russia

E-mail: zuev@imces.ru

Abstract. A technique for automated cloud identification by analyzing measurable direct and diffuse radiation is proposed.

1. Introduction

Having determined the form of a cloud cover, one can get an idea of the processes leading to its formation in the atmosphere and predict their further development. Particular attention should be paid to convective cloudiness, which is a powerful carrier of atmospheric moisture and a regulator of radiation balance and, due to its proximity to the Earth's surface and high frequency in the warm time of the year, has a great influence on the formation of weather and climate and is a major source of hazardous meteorological and emergency situations. Satellite observations for various reasons cannot always provide information on the state of the lower layers of the atmosphere in which such cloudiness is formed and, therefore, ground observations are of great importance in this case. Regular visual observations of basic parameters of cloudiness (shape, amount and cloud base height) are carried out at weather stations 8 times a day, which is clearly not enough, because low clouds are very dynamic formations and can change the state of the sky from clear to completely covered by clouds within 20-30 minutes.

At present, as part of a modernization of the actinometric network, weather stations are being equipped with automatic actinometric complexes (AACs), which allow us to obtain continuous series of measurements of major elements of the radiation mode which directly depend on the state of the atmosphere and primarily on the shape and amount of cloudiness. With series of measurements of the direct S and diffuse D irradiances, it is possible, with some assumptions, to restore the state of the sky that caused a particular combination of the S and D values on the Earth's surface that are naturally and conveniently used as initial parameters.

2. Theory

The standard methodology to determine cloud shapes is based on ae morphological classification which makes it possible to define the shape of clouds by their appearance [1]. The use of such classification in automated systems is difficult, primarily, because for this purpose, in addition to a matrix photodetector for direct monitoring of the sky condition, considerable computing and, hence, energy resources are required to implement complex and time-consuming algorithms for image processing and image recognition. In contrast, for a broad automation of the cloud shape determination process, inexpensive systems that can classify cloudiness by other (non-morphological) features, for example, actinometrical features, are needed. Table 1 shows typical and usual distinctive features of basic cloud shapes [1], according to which all clouds may be subdivided with respect to their optical

Content from this work may be used under the terms of the Creative Commons Attribution 3.0 licence. Any further distribution of this work must maintain attribution to the author(s) and the title of the work, journal citation and DOI.
Published under licence by IOP Publishing Ltd

thickness into three classes: transparent, semi-transparent, and non-transparent ones. Within each of the classes, three groups of clouds may be defined according to the degree of influence on direct irradiance and four groups of clouds according to their effects on diffuse irradiance.

Table 1. Typical and usual features of basic cloud types.

Cloud optical thickness	Cloud forms	Decrease of direct irradiance S			Change of diffuse irradiance D			
					increase			
		weak	significant	strong	weak	significant	strong	decrease
Transparent clouds	*Ci*	x			x			
	Cc	x			x			
	Cs	x			x			
Semi-transparent clouds	*Ac*		x			x		
	As			x		x		
Non-transparent clouds	*As*			x			x	
	Ns			x				x
	Sc			x		x		
	St		x	x		x	x	
	Cu			x		x	x	
	Cb			x	x			x

Thus, each type of cloudiness can be described by specific, unique amplitude and temporal actinometrical features. The amplitude features depend on the effect this particular type of cloudiness has on direct and diffuse irradiances. These are described by the transmission ratio of direct irradiance $C_S = \frac{\overline{S}}{S_0}$ and the ratio of change of diffuse irradiance $C_D = \frac{\overline{D}}{D_0}$, where $\overline{S}$ and $\overline{D}$ are mean values measured in some time period, and S_0 and D_0 are the values of direct and diffuse irradiances in clear sky and at the same Sun altitude. These ratios allow us to probe the clouds not only in terms of the throughput, but also in terms of their ability to re-emit solar radiation. Such an approach was proposed in [2], where the authors only use the transmission ratio of direct irradiance C_S, that is, the cloud is examined only to transmit direct radiation. Since clouds are not stationary and always in motion and development, they form their own temporal actinometrical features which can be described by the coefficients of variation of direct $V_S = \frac{\sigma}{\overline{S}}$ and diffuse $V_D = \frac{\sigma}{\overline{D}}$ irradiance, characterizing the homogeneity of cloud optical thickness and its speed. One can find such an approach in [3], where, as an amplitude feature, the average value of the total irradiance $Q = S' + D$ scaled to a value of 1400 $W \cdot m^{-2}$ and, as a temporal feature, the standard deviation in a 21-minute period are used. The distinctive feature of cumulus cloudiness, for example, is its ability to change the solar disk type with some frequency from a completely open ($S = S_0$) to a completely shadowed ($S = 0$), and also to significantly increase the diffuse irradiance D (up to 3-3.5 times). In this case there may be situations when a reduction in the direct irradiance, when a cumulus cloud is covering the Sun disk, will be considerably compensated by an increase in the diffuse irradiance from the same cloud, thus making the total irradiance decrease insignificant, which is typical not only for cumulus, but also for altocumulus clouds. This uncertainty in the identification of cloud types can be avoided by using separate analyses of the direct *S* and diffuse *D* irradiances.

The technique to define cloud types using an optical thickness classification proposed below involves measuring the direct *S* and diffuse *D* irradiances within a certain period of time, determining their mean values, calculating the coefficients $C_S V_S C_D V_D$, and comparing them with reference values for some shapes and types of clouds. For cumulus cloudiness the technique assumes determining the mean values of these coefficients for each cloud amount.

3. Raw data and measurement results

To apply this technique for automated cloud detecting in practice, it is necessary to solve two problems. The first is to choose (or create) a model of diurnal variation of the direct S_0 and diffuse D_0 clear-sky irradiances for the measurement point, season, and Sun altitude considered. In [3], to model the diurnal total clear-sky irradiance Q_0, a complex analytical expression is used which includes both empirical and measured coefficients of absorption and diffusion in atmospheric gases, atmospheric pressure, and dew point. In our case, interpolated average long-term values for clear sky of the warm period of the year for a given point are used to model the clear-sky diurnal variation of the S_0 and D_0 irradiances [4].

The second problem is determining the mean values of the coefficients $C_S V_S C_D V_D$ for each type of clouds and separately for each cumulus cloud amount, which demands continuous synchronous series of instrumental measurements of the direct and diffuse irradiances and of visual observations of cloud shapes for comparison. As the series, one-minute measurements of the direct and diffuse irradiances from the AAC of the Ogurtsovo meteorological station (Novosibirsk, WMO 29638) for May-August 2016 were used [5]. The mean values of $\overline{S}$ and $\overline{D}$ were calculated for 20-minute periods with a single shape of cloudiness on with clear sky. The shape and cloud amount were identified from panoramic images of the sky from a television meter of the cloud parameters [6] set up near the AAC. Figure 1 shows an example of processing of a panoramic image when there were only cumulus clouds in the sky.

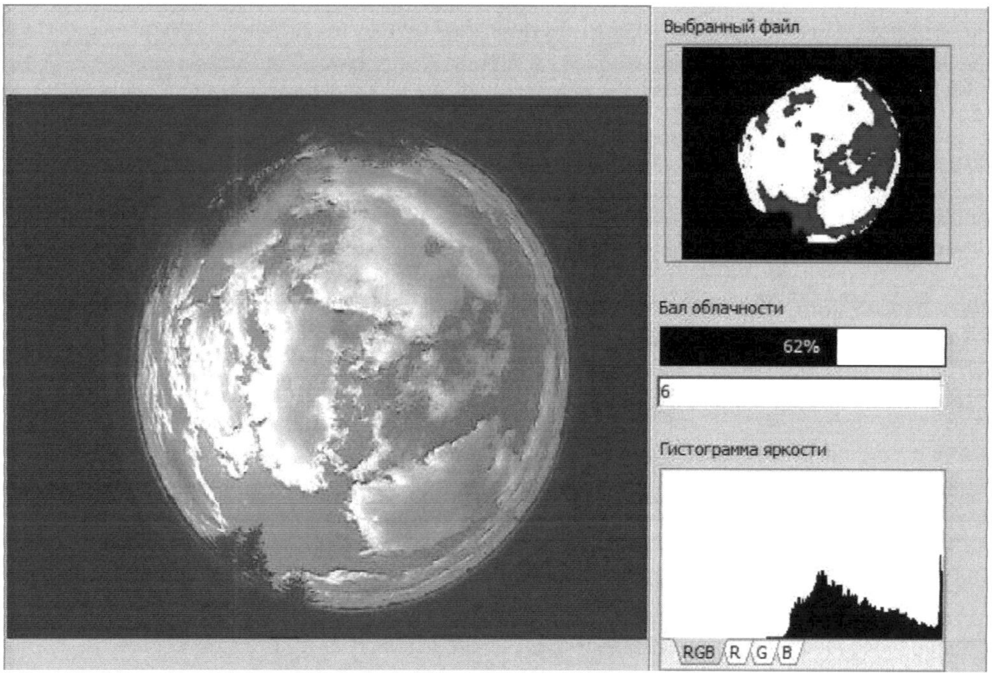

Figure 1. Original and processed all-sky image.

The thus obtained mean values of the coefficients $C_S V_S C_D V_D$ for the main shapes and types of clouds are presented in Table 2, where the clear sky condition is described similarly to the cloud condition.

Table 2. Average coefficients of basic cloud types.

Cloud optical thickness	Cloud forms (sky condition)	Average coefficients			
		C_S	V_S	C_D	V_D
Full transmission	Clear	1.0	0.01	0.9	0.04
Transparent clouds	***Ci***. ***Cc***. *Ac trans.*	0.7	0.7	2.2	0.06
	Cs	0.7	0.5	2.1	0.02
Semi-transparent clouds	***Ac***. *Ci sp. Sc trans.*	0.5	0.7	2.6	0.10
	As	0.7	0.5	3.8	0.04
Non-transparent clouds	***Sc***. *Ac op.. Ac lent.*	0.1	0.89	3.2	0.15
	St. *As neb.op.. As und. op.*	0.3	0.5	2.1	0.02
	Cu. *Ci ing.*	0.6	0.62	1.8	0.18
	Cb. ***Ns***	0	0.9	1.5	0.21

It can be seen that a small coefficient of variation of the diffuse irradiance ($V_D \leq 0.05$) is characteristic for clear sky and for stratiform clouds (*Cs*, *As*, *St*), which is explained by the homogeneity of the optical thickness of the cloud layer (or the atmosphere in the case of clear sky) and by the low speed of its motion across the sky. A large coefficient of variation of the direct irradiance ($V_S > 0.6$) is characteristic for cumuliform clouds (*Cc*, *Ac*, *Sc*, *Cu*, *Cb*), which is explained by the "trigger" condition of the solar disk, which can change its state from shadowed to open at some frequency depending on the cloud amount and velocity.

Table 3. Average coefficients of cumulus clouds.

Cumulus cloud amount *n*	C_S	V_S	C_D	V_D	Number of measurements *N*
0	1.0	0.01	0.8	0.04	32
1	0.9	0.13	1.1	0.05	94
2	0.9	0.19	1.2	0.10	331
3	0.8	0.31	1.3	0.12	919
4	0.7	0.48	1.5	0.13	1082
5	0.6	0.74	1.8	0.13	907
6	0.5	0.94	2.0	0.13	840
7	0.3	1.17	2.1	0.14	835
8	0.2	1.68	2.0	0.16	528
9	0.1	1.87	1.9	0.19	200

In view of the fact that under cumulus cloudiness the range of values of the coefficients $C_S V_S C_D V_D$ is quite large, it is advisable to divide the interval of their values into several domains for each cloud amount. To calculate the average values of the coefficients, a 20-minute measurement interval was also used, when there was only a certain amount of cumulus cloudiness in the sky. The average values of the coefficients $C_S V_S C_D V_D$ obtained for various cumulus cloud amounts are given in Table 3.

It is well-known that the magnitude of the diffuse irradiance depends to a large extent on the shape and amount of cloudiness. With clear sky it depends on the Sun altitude angle, atmospheric conditions and surface albedo and $D = D_0$. High-level clouds typically cause an insignificant increase in the diffuse irradiance compared to its value for clear sky and $D = D_0 + D_h$ ($D_h > 0$). Middle-level clouds, depending on their type, can cause both large and small increase in the diffuse irradiance (for example, *Ac trans.* or *Ac op.*, accordingly) and $D = D_0 + D_m$, where $D_m = (1.6 \div 5)D_0$. Low-level clouds and clouds of vertical development can both increase and decrease (in case of rain clouds) the diffuse irradiance. Located at low altitudes (the cloud base height under 1500 m), such clouds cover clear-sky areas and have a significant effect on D_0, which must be taken into account. Therefore, for low-level clouds $D = (1 - b)D_0 + D_l$, where b is the cloud amount in fractions of units. In general, the cloud amount can be considered as a function of the diffuse irradiance $b = f(D)$. With average values of the C_D (from Table 3), it becomes possible to estimate the cumulus cloud amount [7]. This dependence is shown in Figure 2.

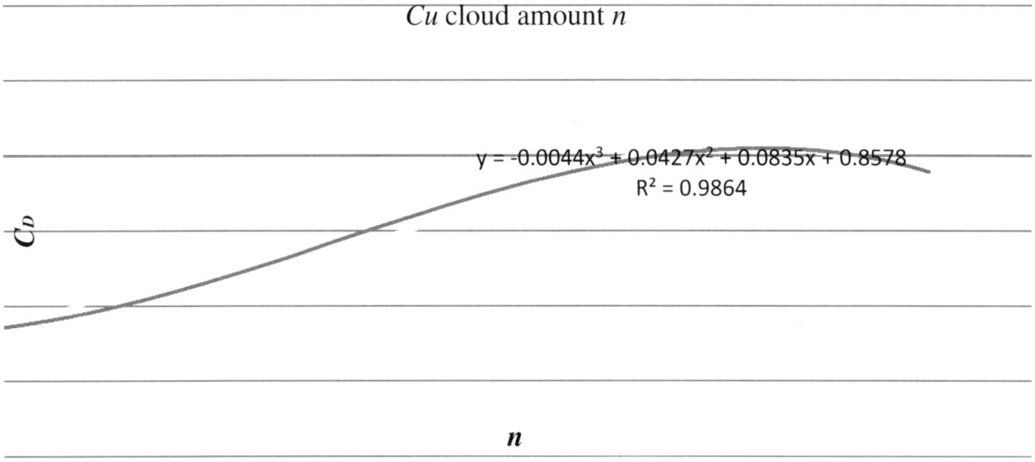

Figure 2. Coefficient C_D versus cumulus cloud amount.

4. Conclusions

Clouds are a natural regulator of the solar radiation coming to the Earth's surface. All major types of cloudiness can be described by their unique amplitude and temporal actinometrical features, which characterize their ability to transmit and re-emit solar radiation. Cumulus cloudiness is of great significance for the process of weather and climate formation of a territory, since it is dynamically unstable and can carry large amounts of water over long distances both horizontally and vertically. Besides, this type of cloudiness is the basis for the formation of cumulonimbus clouds, which are most dangerous in terms of emergency situations in the sky and on the Earth. The paper presented a technique to determine the cloud shape by its amplitude and temporal actinometrical features based on an analysis of continuous series of data for the direct and diffuse irradiances. This approach allows us to formalize and automate the process, which in most cases is carried out visually. Also, making separate analyses of the constituents of the total irradiance allows determining not only the type of cloudiness, but also its amount by the magnitude of the diffuse irradiance.

Acknowledgments
The research was carried out with financial support of SB RAS (project of fundamental research no. IX.138.2.5) and the Ministry of Education and Science of the Russian Federation (project no. 5.3279.2017 / PCh).

References
[1] Bespalov D P *et al* 2011 *Cloud atlas* ed L K Surigina (St. Peterburg: D'ART) p 248
[2] Martínez-Chico M , Batlles F J and Bosch J L 2011 Cloud classification in a mediterranean location using radiation data and sky images *Energy* **36** 4055–62
[3] Duchon C E and O'Malley M S 1999 Estimating cloud type from pyranometer observations *J. Appl. Meteorol.* **38** 132–41
[4] *Scientific and Applied Climate Handbook. 3. Long-term data series. Parts 1-6. Issue 20. Tomsk, Novosibirsk, Kemerovo Region, Altai Territory* 1993 (St. Peterburg: Gidrometeoizdat) p 718
[5] Zuev S V , Krasnenko N P and Kartashova E S 2016 Actinometrical features of basic cloud types *IOP Conference Series: Earth and Environmental Science* **1** (48) 012027
[6] Zuev S V, Krasnenko N P and Levikin V A 2014 TV instrument for measuring of cloud characteristics *Proceedings of TUSUR* **1** (31) 54–9
[7] Zuev S V 2016 Using diffuse solar radiation for estimate of *Cu* cloud amount 2016 *Proc. SPIE* 10035 100354B

Integration of modern statistical tools for the analysis of climate extremes into the web-GIS "CLIMATE"

A A Ryazanova[1,2], I G Okladnikov[1,2,3] and E P Gordov[1,2,3]

[1]Institute of Monitoring of Climatic and Ecological Systems SB RAS, Tomsk, Russia
[2]V.E. Zuev Institute of Atmospheric Optics SB RAS, Tomsk, Russia
[3]Institute of Computational Technologies, Tomsk Branch, Tomsk, Russia

E-mail: raa@scert.ru, igor.okladnikov@gmail.com, gordov@scert.ru

Abstract. The frequency of occurrence and magnitude of precipitation and temperature extreme events show positive trends in several geographical regions. These events must be analyzed and studied in order to better understand their impact on the environment, predict their occurrences, and mitigate their effects. For this purpose, we augmented web-GIS called "CLIMATE" to include a dedicated statistical package developed in the R language. The web-GIS "CLIMATE" is a software platform for cloud storage processing and visualization of distributed archives of spatial datasets. It is based on a combined use of web and GIS technologies with reliable procedures for searching, extracting, processing, and visualizing the spatial data archives. The system provides a set of thematic online tools for the complex analysis of current and future climate changes and their effects on the environment. The package includes new powerful methods of time-dependent statistics of extremes, quantile regression and copula approach for the detailed analysis of various climate extreme events. Specifically, the very promising copula approach allows obtaining the structural connections between the extremes and the various environmental characteristics. The new statistical methods integrated into the web-GIS "CLIMATE" can significantly facilitate and accelerate the complex analysis of climate extremes using only a desktop PC connected to the Internet.

1. Introduction

Climate extremes may have significant effects on natural and artificial terrestrial ecosystems. However, their impact is complex, depends on numerous factors, and has not yet been fully understood so far. A detailed knowledge of the statistics of climate extremes and spatio-temporal patterns of the essential climate variables causing such events is important for estimating the potential damages and, hence, sustainable planning and managing the ecosystem-related resources. It is well-known that since 1970 both the average global temperature and total precipitation amount are increasing [1]. Also, the frequency of occurrence and magnitude of the events directly related to the extreme values of these characteristics (showers, hurricanes, droughts, and heat waves) have clear growth trends [1]. Detailed analyses and understanding of their impact on the environment are needed for the forecasting of such extremes and mitigation of their effects. Such studies must employ modern statistical methods for the analysis of extreme values [2-6]. To characterize climate extremes, data from various sources (meteorological stations, reanalysis products, remote sensing data) covering various spatial and temporal ranges and scales are used. To simplify the use of extreme value statistics

Content from this work may be used under the terms of the Creative Commons Attribution 3.0 licence. Any further distribution of this work must maintain attribution to the author(s) and the title of the work, journal citation and DOI.
Published under licence by IOP Publishing Ltd

tools and handle large volumes of such data, we augmented thematic web-GIS called "CLIMATE" developed for cloud storage processing of geophysical datasets [7, 8]. The integration of these tools into the web-GIS "CLIMATE" and some results are presented in this paper.

2. Methods and approaches

The system "CLIMATE", based on web- and GIS-techniques, is part of a hardware and software complex for cloud storage analysis of climate data. This complex includes various climatological and meteorological datasets, as well as dedicated interactive tools for their search, sampling, processing, and visualization. Utilization of this system significantly simplifies and accelerates the handling of huge volumes of spatial climate datasets, allowing users without specific IT knowledge to perform remotely processing and analysis of data using only a modern desktop PC connected to the Internet. The web-GIS allows calculating the basic and complex statistical properties of time series of meteorological and climatological characteristics [7-9], including average, minimum, maximum, dispersion, and a set of basic WMO climate change and extreme indices (http://etccdi.pacificclimate.org/) and hydrothermal coefficients.

Powerful, more sophisticated alternatives accounting for trends in extremes (like time-dependent extreme value statistics [10, 11] or quantile regression [12, 13]) are available but have not yet been systematically exploited. Since complex climate extremes (e.g., of temperature and precipitation) are of particular relevance for ecosystems, the well-known copula approach for describing multivariate probability distributions [14, 15] provides a promising technique, which has also not yet been used in this context. These advanced methods for climate data analysis significantly extend the current functionality of the system "CLIMATE" and seem very promising for climate research applications.

2.1. Time-dependent statistics of extremes

A statistical description of extreme events, such as extreme precipitation or temperature, can be done using the concepts of extreme value statistics (EVS). Unlike in the early stages of EVS [16], we can now take advantage of well-developed approaches to describe univariate extremes with non-stationary, i.e. time-dependent, probability distribution functions (PDFs). EVS provides two major routes to a probabilistic description of extremes: *the block maxima approach* to describe the probability distribution of the intensity of maxima of blocks (e.g., monthly or annual maximum of a meteorological variable) with a generalized extreme value distribution (GEV), and the *peak-over-threshold approach*, modelling excesses over a threshold with a generalized Pareto distribution. For example, you can use the block maxima approach and a time-dependent model to describe the seasonal variability of precipitation maxima [10]. A well-written introduction to EVS covering both the routes and non-stationary concepts is given by Coles [17].

A software implementation of EVS in the R language is represented by the class "fevd" from the software package "extRemes" [18, 19]. The input arguments for an instance of this class depend on the selected approach. If the *block maxima approach* is chosen, the maximum values of the analyzed meteorological characteristics for a selected time range (e.g., month or year) are given. If the *peak-over-threshold approach* is chosen, daily or maximum values for selected time range values are given, but an additional input argument "threshold" must be given also. Input argument "type" allows us to select PDF (by default GEV is used). The input argument "method" allows us to select estimation parameters (by default "Maximum Likelihood Estimation" – MLE is used). The arguments "location.fun", "scale.fun", "shape.fun", and "threshold.fun" allow us to set time dependencies of PDF parameters. The result of a calculation is a set of estimated parameters of a given PDF (in the case of GEV it is a vector with location, scale, and shape parameters). A more detailed description can be found in [18].

The PDF location, scale, and shape parameters by themselves provide little information on the probability of occurrence of extreme events. A physically more meaningful and also more relevant quantity for risk assessment is the probability of an observed variable exceeding a certain level. These levels can be calculated from a parameterized GEV and are frequently expressed as return levels r_T for a certain return period T. r_T is defined as the level which is exceeded on average every T blocks, i.e., with probability $^1/_T$ [10]. r_T is calculated using «return.levels» class with a list of class «fevd» instances as the major argument. There are ad-on arguments: «return.period», a numeric vector of desired return periods and «qcov», a numeric matrix with rows the same length as q and columns equal to the number of parameters (+ 1 for the threshold, if a POT model). This gives any covariate values for a nonstationary model and is not used in the stationary case. «qcov» is used to obtain effective return levels.

As a result, an object of class "return.level" is returned, which is either a numeric vector (stationary models) of length equal to the «return.period» argument giving the return levels or a matrix of a dimension equal to either n by np or q by np, where n is the length of the data used to fit the model and np is the number of return periods, and q is the number of rows of «qcov», if supplied. The returned value also includes useful attributes describing how the return levels are to be estimated. A detailed description of the «return.levels» class can be found in [18].

2.2. *Quantile regression*

The analysis of trends in meteorological observations is one of the most common activities in climate change studies. Quantile regression provides a well-defined statistical framework for estimating the rate of change not only in the mean as in ordinary regression, but in all parts of the data distribution. It was first introduced in econometrics by [5]. Now quantile regression is used in various geoscience contexts [12, 20-24].

Given a random variable Y with a cumulative continuous distribution function $F_Y(y)$, the quantile is defined as the value $Q_Y(\tau)$ such that $P[Y \leq Q_Y(\tau)] = \tau, 0 \leq \tau \leq 1$. The quantile function $Q_Y(\tau)$ is defined from the cumulative distribution function $F_Y(y)$ as $Q_Y(\tau) = F_Y^{-1}(\tau)$. Then, considering the conditional distribution of Y given $X = x$, the conditional quantile function $Q_{Y|X}(\tau; x)$ verifies $P\left[Y \leq Q_{Y|X}(\tau; x)|X = x\right] = \tau$. Whereas ordinary regression is based on the conditional mean function $E[Y|X = x]$ and minimization of the respective residuals, quantile regression is based on the conditional quantile function and minimization of the sum of asymmetrically weighted absolute residuals $\sum_{i=1} \rho(\tau)|y_i - Q_{Y|X}(\tau; x = x_i)|$, where ρ is the tilted absolute value function.

Another advantage of quantile regression is that it is non-parametric, i.e. the distribution of the investigated variable is not assumed to belong to any parametric population. Moreover, quantile regression is a robust method and weakly affected by the outliers and peak (extreme) values. Further details can be found in [25, 26].

A software implementation of quantile regression in the R language is represented by the class "qr" from the software package "quantreg" [27]. To calculate the quantile trends, the input argument "formula" allows us to set a formula describing a quantile model (e.g., in the simplest case we can set a linear time dependence of the meteorological values; in a more complicated case the model can be extended with seasonal components). The quantile values of interest are set between 0 and 1 in the input vector "tau". The results are returned as a vector of quantile regression model estimated parameters. A detailed description of "qr" class can be found in [27].

2.3. *Copula approach*

Due to a possible nonlinear dependence between the variables, the probability distributions of multivariate random variables are generally more complex than their univariate counterparts. One of the promising approaches to this problem is the use of copulas. It has become popular in recent years, especially in econometrics, finance, risk management, or insurance. The copula approach is a simple and straightforward method to find parametric descriptions of multivariate non-normally distributed random variables. Although it is a fast growing field of statistics, it is still little known in climate research, where often non-normally distributed random variables like precipitation, wind speed, cloud cover, humidity, etc. are involved. An advantage of this approach is the ability to catch various covariance structures while keeping proper parametric descriptions of the margins.

IOP Conf. Series: Earth and Environmental Science **96** (2017) 012014 doi:10.1088/1755-1315/96/1/012014

Unlike the univariate extreme value theory, here the possible limiting distribution functions of extreme values cannot be captured by a finite-dimensional parametric family of functions. The study of multivariate extremes splits into descriptions of the marginal distribution and the dependence structure. In order to characterize the limit behavior of multivariate extremes, it has been shown that weak convergence of the multivariate distribution is equivalent to weak convergence of the marginals as well as the copula function, provided that the marginals are continuous [3]. This is the reason why the copula approach is very popular in modeling multivariate extremes [28].

Possible applications of the copula approach include different typical selections of pairs of variables, like one meteorological characteristic at different locations or different meteorological characteristics at the same location, as far as they can be expressed in the copula framework [14]. There is no general procedure for selecting the copula class.

A software realization of the copula approach technique is represented in the R language as the dedicated stand-alone software package "copula" [29, 30].

3. Results

The above-described software tools for statistical analysis of time series of spatial climatological datasets have been integrated into the web-GIS "CLIMATE", thus allowing analysis of current and future extreme climate events. The algorithms from the packages "extRemes" [18, 19], "quantreg" [27], and "copula" [29, 30] implementing time-dependent statistics of extremes, quantile regression and the copula approach were used as the basis for these tools. These packages are written in the R language (https://www.r-project.org/, [31]) and provide a flexible API for the integration into a third-party software. The computing core of the web-GIS "CLIMATE" is written in the GNU Data Language (GDL, http://gnudatalanguage.sourceforge.net/), and currently there is lack of universal wrappers providing the execution of R procedures inside a GDL program. To overcome this obstacle, the fact that GDL has built-in Python (https://www.python.org/) interface was used and special software adapters connecting GDL and R were developed in Python. These adapters utilize the dedicated Python-package RPy2 [32] and provide a two-way data and instruction transfer between the web-GIS computing core and procedures in R packages (Figure 1). As a result, the computing core acquired the ability to call new procedures implemented in the R language to perform complex statistical analysis of climate data.

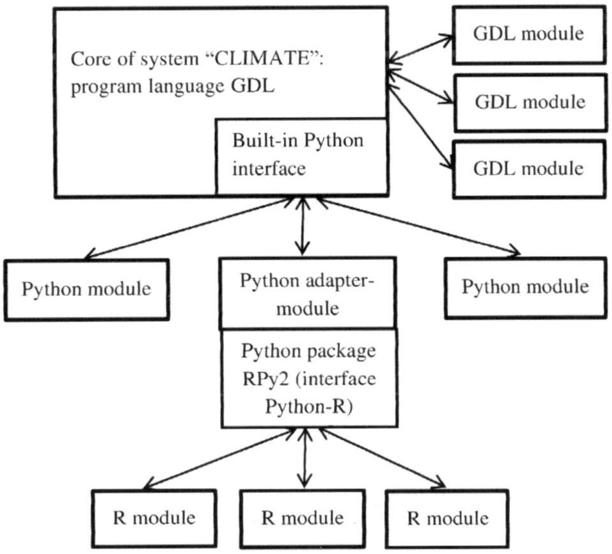

Figure 1. Diagram of interactions between computing core and R-modules.

IOP Conf. Series: Earth and Environmental Science **96** (2017) 012014 doi:10.1088/1755-1315/96/1/012014

To provide an intuitively clear and user friendly graphical interface of the online tools as well as to provide effective processing and data handling, the corresponding metadata describing new processing routines were inserted during the integration process into a web-GIS dedicated metadata database. This database [33] contains spatial and temporal characteristics of the available spatial datasets, their physical locations, and the parameters of software procedures for data analysis and visualization. Thanks to this database it has become possible to extend the processing abilities of the web-GIS and to edit the GUI content without any modification of the source codes of the web-GIS components. All new abilities have become available to the users immediately after the metadata insertion, thus providing a rapid and sustainable development of the web-GIS "CLIMATE".

To illustrate the extended web-GIS "CLIMATE"'s functionality, the software module utilizing the package "extRemes" was used to calculate 100-year return levels of maximum precipitation conditioned on the July month on ECMWF ERA Interim [34] (Figure 2a) and on APHRODITE JMA [35] data (Figure 2b) for Southern Siberia (52.5-60° N, 75-95° E).

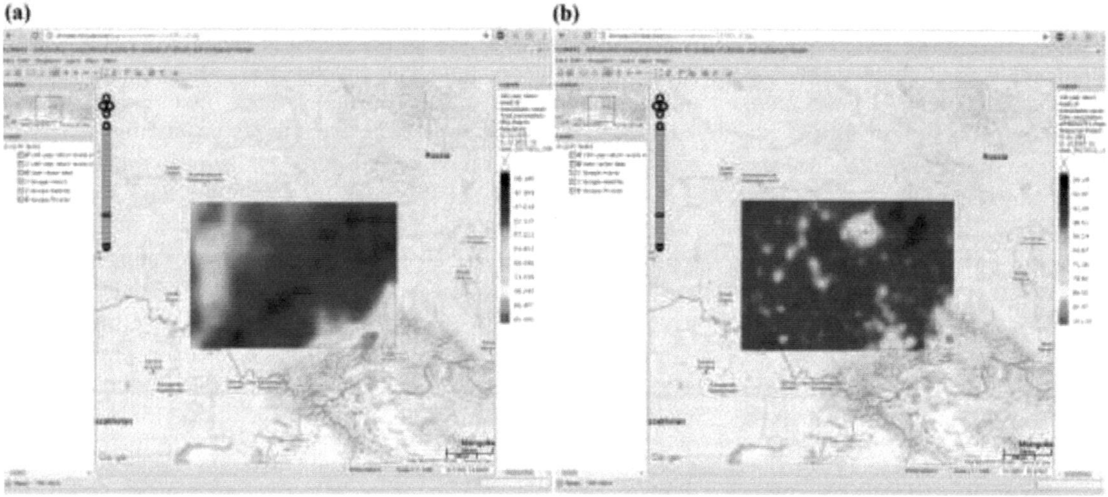

Figure 2. Screen shots of obtained cartographic layers for 100-year return levels of maximum precipitation conditioned on the July month of their occurrence for Southern Siberia: a) on ECMWF ERA Interim data, 0.75x0.75 horizontal grid, 1979-2012, b) on APHRODITE JMA data, 0.25x0.25 horizontal grid, 1951-2007.

The results obtained are in good agreement with the information available in the literature [10].

4. Conclusions

The integration of modern statistical packages for extreme analysis developed in the R language into the web-GIS "CLIMATE" significantly extended its functional abilities. Thanks to it, new powerful climate data analysis techniques, such as time-dependent statistics of extremes, quantile regression and copula approach, have become available to the users of the web-GIS "CLIMATE". These techniques allow performing detailed analysis of various extreme climate events, to estimate the magnitude of their impact, and to discover the structural interconnections between the extremes and the environment characteristics.

A practical application of the earlier developed metadata database showed that its usage improves the scalability and flexibility of computations and facilitates new processing procedures to the web-GIS "CLIMATE".

The results obtained show that the new interactive tools for the extreme climate events analysis can be useful for decision makers and specialists working in affiliated sciences, to make socio-economic impact assessment, ecological impact assessment, adaptation strategies, science policy administration, and other climate related activities. These tools can provide reliable climate related characteristics required for studies of economic, political, and social consequences of global climate change at the regional level.

Acknowledgements
This work was supported by SB RAS under Basic Research Program no. IX.138.2.1.

References

[1] Stocker T F, Qin D, Plattner G-K, Tignor M, Allen S K, Boschung J, Nauels A, Xia Y, Bex V and Midgley P M 2013 *Climate Change 2013*: *The Physical Science Basis. Contribution of Working Group I to the Fifth Assessment Report of the Intergovernmental Panel on Climate Change* (Cambridge: Cambridge University Press and NY, USA) p 1535
[2] Friederichs P 2007 *An Introduction to Extreme Value Theory* (Bonn: Meteorological Institute University of Bonn, COPS, Summer School)
[3] Beirlant J, Goegebeur Y, Segers H and Teugels J 2004 *Statistics of Extremes: Theory and Applications* (Wiley Series in Probability and Statistics)
[4] Embrechts P, Kluppelberger C and Mikosch T 1997 *Modelling Extremal Events for Insurance and Fincance* (Berlin: Springer)
[5] Koenker R and Basset 1978 Regression quantiles *Econometrica* **46** 33–50
[6] Nelsen R 2006 *An Introduction to Copulas* (New York: Springer, 2nd edn)
[7] Gordov E P, Okladnikov I G, Titov A G, Bogomolov V Yu, Shulgina T M and Genina E Yu 2012 Geo-information system for investigation of regional climatic changes and first results obtained *Atmos. Ocean. Opt.* **25** (2) 137–43
[8] Gordov E P, Shiklomanov A, Okladnikov I G, Prusevich A and Titov A G 2016 Development of Distributed Research Center for analysis of regional climatic and environmental changes *IOP Conf. Series: Earth and Environmental Science* **48** doi:10.1088/1755-1315/48/1/012033
[9] Riazanova A A, Voropay N N, Okladnikov I G, Gordov E P 2016 Development of computational module of regional aridity for web-GIS "Climate" *IOP Conf. Series: Earth and Environmental Science* **48** doi:10.1088/1755-1315/48/1/012032
[10] Rust H, Maraun D and Osborn T J 2009 Modelling seasonality in extreme precipitation *Europ Phys J ST* **174** 99–111
[11] Katz R W, Parlange M B and Naveau P 2002 Statistics of extremes in hydrology *Advances in Water Resources* **25** 1287–1304
[12] Barbosa S M, Scotto M G and Alonso A M 2011 Summarising changes in air temperature over Central Europe byquantile regression and clustering *Nat Hazards Earth Syst Sci* **11** 27–3233
[13] Sterin A M and Timofeev A A 2014 Specific features of estimates of surface air temperature trends in the Russian Federation obtained using quantile regression *Proceedings of RIHMI-WDC* **178**
[14] Schölzel C and Friederichs P 2008 Multivariate non-normally distributed random variables in climate research – introduction to the copula approach *Nonlin Proc Geophys* **15** 761–72
[15] Salvadori G and De Michele C 2004 Frequency analysis via copulas: Theoretical aspects and applications to hydrological events *Water resources research* **40**
[16] Fisher R A and Tippett L H C 1928 Limiting Forms of the Frequency Distribution of the Largest or Smallest Members of a Sample *Proc. Cambridge Phil. Soc.* **24** 180–90
[17] Coles S G 2001 *An Introduction to Statistical Modelling of Extreme Values* (London: Springer)
[18] Gilleland E 2016 Package "extRemes" The Comprehensive R Archive Network (CRAN) https://cran.r-project.org/web/packages/extRemes/extRemes.pdf
[19] Gilleland E and Katz R W 2016 extRemes 2.0: An Extreme Value Analysis Package in R *Journal of Statistical Software* **72** (8) doi: 10.18637/jss.v072.i08

[20] Koenker R and Schorfheide F 1994 Quantile spline models for global temperature change *Climatic Change* **28** (4) 395–404
[21] Cade B and Noon B 2003 A Gentle introduction to quantile regression for ecologists *Front. Ecol. Environ.* **1** 412–20
[22] Baur D, Saisana M and Schulze N 2004 Modelling the effects of meteorological variables on ozone concentration – a quantile regression approach *Atmos. Environ.* **38** 4689–99
[23] Elsner J B, Kossin J P and Jagger T H 2008 The increasing intensity of the strongest tropical cyclones *Nature* **455** 92–5
[24] Barbosa S M 2008 Quantile trends in Baltic sea-level *Geophys. Res. Lett.* **35** L22704 doi:10.1029/2008GL035182
[25] Koenker R and Hallock K 2001 Quantile Regression *J. Economic Perspect.* **15** 143–56
[26] Koenker R 2005 *Quantile regression* (New York: Cambridge University Press)
[27] Koenker R, Portnoy S, Tian P, Zeileis A, Grosjean P and Ripley B D 2017 Package "quantreg" The Comprehensive R Archive Network (CRAN) https://cran.r-project.org/web/packages/quantreg/quantreg.pdf
[28] Renard B and Lang M 2007 Use of a Gaussian copula for multivariate extreme value analysis: Some case studies in hydrology *Adv. Water Resour.* **30** 897–912
[29] Hofert M, Kojadinovic I, Maechler M, Yan Y 2017 Package "Copula" The Comprehensive R Archive Network (CRAN) ftp://cran.r-project.org/pub/R/web/packages/copula/copula.pdf
[30] Yan J 2007 Enjoy the Joy of Copulas: With a Package copula *Journal of Statistical Software* **21** 4
[31] Gilleland E 2011 *Using R to Analyze Extremes* National Center for Atmospheric Research (Boulder, Colorado, U.S.A)
[32] Documentation for RPy2 http://rpy2.readthedocs.io/en/version_2.8.x/index.html
[33] Okladnikov I G, Gordov E P and Titov A G 2016 Development of climate data storage and processing model *IOP Conf. Series: Earth and Environmental Science* **48** doi:10.1088/1755-1315/48/1/012030
[34] Dee D P *et al.* 2011 The ERA-Interim reanalysis: configuration and performance of the data assimilation system Quarterly *Journal of the Royal Meteorological Society* **137** (656) Part A 553–97
[35] APHRODITE JMA, http://www.chikyu.ac.jp/precip/data/APHRO_V1003R1_readme.txt

IOP Conf. Series: Earth and Environmental Science **96** (2017) 012015 doi:10.1088/1755-1315/96/1/012015

Droughts and Excessive Moisture Events in Southern Siberia in the Late XXth - Early XXIst Centuries

A A Ryazanova [1] and N N Voropay [1,2]

[1] Institute of Monitoring of Climatic and Ecological Systems SB RAS, Tomsk
[2] V.B. Sochava Institute of Geography SB RAS, Irkutsk

E-mail: raa@scert.ru, voropay_nn@mail.ru

Abstract. In recent years much research has been devoted to global and regional climate changes. Special attention was paid to climate extremes, such as droughts and excessive moisture events. In this study the moisture and aridity of Southern Siberia are estimated using web-GIS called "CLIMATE". The system "CLIMATE" is part of a hardware and software cloud storage complex for data analysis of various climatic data sets, with algorithms for searching, extracting, processing, and visualizing the data. The ECMWF ERA-Interim reanalysis data for Southern Siberia (50-65°N, 60-120°E) from 1979 to 2010 with a grid cell of 0.75×0.75° is used. Some hydrothermal conditions are estimated using the so-called Ped index (S_i), which is a normalized indicator of the ratio of air temperature to precipitation. The mountain regions of Eastern Siberia are becoming more and more arid each month during the last 30 years. In Western Siberia, aridity increases in May and decreases in June, in the other months positive and negative trends are found. The greatest differences between the trends of the aridity index (S_i), air temperature, and precipitation are observed in July.

1. Introduction

The global and regional climate changes have been intensively discussed in the recent years by scientific community [1, 2]. Monthly and annual air temperature and atmospheric precipitation are usually analyzed. Changes in extreme values of air temperature and precipitation are analyzed rarely. Extreme climate parameters are associated with drought and excessive moisture periods.

A drought is a complex natural phenomenon with strong regional anomalies of temperature and humidification [3]. A drought is defined as a period in which humidity is below a statistical average expected for that location. It is part of natural variability of climate and weather. Due to the complex structure of a drought, scientists are usually limited to studying one of its aspects. Depending on the purpose of study, they investigate atmospheric drought or soil drought only. Unfavorable weather or climatic conditions significantly reduce the agricultural potential of climate. Active heating and relative air desiccation occur at an atmospheric drought. At the same time, there is no precipitation and large incoming solar radiation.

Various approaches to the definition of drought are widely discussed depending on characteristic types and conditions of drought formation. Some representative aridity indices were developed [4-7], catastrophic droughts were analyzed [8-10], and a catalog of droughts was composed [11].

Content from this work may be used under the terms of the Creative Commons Attribution 3.0 licence. Any further distribution of this work must maintain attribution to the author(s) and the title of the work, journal citation and DOI.
Published under licence by IOP Publishing Ltd

There is no clearly expressed periodicity in droughts or a certain trend in the frequency and intensity of droughts over the territory of Russia in the 20th century. A tendency to increase the frequency of droughts is observed only in some regions [1].

Traditionally, aridity/moisture is assessed in a manual or semi-automatic mode for a restricted area with data from weather stations (observation data). This approach requires significant time for long-term observation data processing and characterizes the aridity and moisture of a territory only in regions where weather stations exist. For vast areas of Siberia and Far East where the weather station data are highly sparse and irregular, this approach significantly reduces the reliability of the obtained results. Using the global retrospective analysis (re-analysis) data causes difficulties too. Large volumes of data (up to several terabytes) require powerful computing complexes for processing and a large storage space, and the dedicated file formats used for storing the data require specific programming skills.

2. Objects, data and methods

In this paper we estimate the aridity/moisture of the territory with the ERA-Interim reanalysis data using a new program module of a previously developed web-GIS "CLIMATE" [12]. The system "CLIMATE" is based on a combined use of web and GIS technologies. It is part of a hardware and software complex for "cloud" storage data analysis using various climatic data sets, as well as algorithms for searching, extracting, processing, and visualizing the data. This system significantly facilitates and accelerates the analysis of big volumes of geospatial data, allowing the researchers to perform complex climate data analysis using desktop PCs with internet connection. At present the system already has many computational modules that allow calculating both the standard statistical characteristics of meteorological values and the extreme indices recommended by the World Meteorological Organization (WMO) Expert Team on Climate Change Detection and Indices (ETCCDI, (http://etccdi.pacificclimate.org/)), as well as the Selyaninov hydrothermal coefficient [13].

Due to the fact that in the studied territory of Southern Siberia (50-65° N, 60-120° E) there is a great variety of landscapes, it was decided to assess the hydrothermal conditions using the so-called Ped index (S_i) [6], which is the dimensionless normalized difference of air temperature and precipitation.

$$S_i = \frac{\Delta T_i}{\sigma_T} - \frac{\Delta P_i}{\sigma_P} \quad , \qquad (1),$$

where i is the number of a certain period, $\Delta T_i = T_i - T_{norm}$ is the temperature anomaly in the i-th period, T_{norm} is the long-term average temperature, σ_T is a standard deviation of temperature. The notations are similar for precipitation (P).

The web-GIS "CLIMATE" was extended by a software module to calculate the Ped hydrothermal index. The module was developed using an Interactive Data Language (IDL 6.0). The algorithm was based on formula (1). The ECMWF (European Centre for Medium-Range Weather Forecasts, UK) ERA-Interim reanalysis data [13] on air temperature and the sum of precipitation at the Earth's surface were used. The Ped hydrothermal index was calculated for the territory of Southern Siberia (50°-65° N, 60°-120° E) from 1979 to 2010, which allows obtaining spatial assessments of aridity/moisture of the territory in contrast to the existing assessments of other authors.

In [13, 15] it was established that the reanalysis and observation data show a slight difference in the air temperature values (the average difference is -0.86°C) and a significant discrepancy (49% on average) in the precipitation. To improve the quality of the precipitation reanalysis data, we use an approach that was presented in a previous study [13]. According to this approach, the daily precipitation reanalysis data on the reanalysis grid are interpolated for the coordinates of weather stations using a previously developed software module for the correction of reanalysis data. 127 weather stations with most complete and homogeneous time series of observations were used. Then the precipitation was corrected using a bilinear regression equation. The correction was performed individually for each year and month of the vegetation period (from May to September) for the study area. The observation data on daily precipitation were taken as a dependent variable, and the reanalysis daily precipitation interpolated for the coordinates of weather stations were taken as an independent variable. The obtained coefficients of bilinear regression were applied to the precipitation reanalysis data. The final step of the calculation and analysis corrected the reanalysis data and was carried out for the hydrothermal Ped index and its components (average monthly temperature and corrected monthly precipitation). This approach allowed us to characterize the hydrothermal conditions not only for the coordinates of weather stations, but also for complicated orographic conditions in areas with a sparse observation network.

3. Results and discussion

The long-term monthly average temperature of the study area changes during the warm period (May-September) from – 1 to + 25 °C of the study area (Figure 1). In general, the latitudinal distribution of monthly average temperature is observed for all months. In Eastern Siberia the orography violates to a certain extent the pattern. The surface elevation varies from 0 (West Siberian Plain) to 4506 m a.s.l. (Belukha mountain, Altai) over the study area. The annual maximum of air temperature is observed in July. The highest air temperatures were recorded in the south-west of the territory (+ 25 °C, Northern Kazakhstan). The July air temperature decreases to +17 °C as a function of latitude. The July air temperature does not exceed +10-12 °C in the mountain regions of Altai and Sayans. The monthly air temperature in the mountains of Transbaikalia (Stanovoy Highland) increases to 13-15 °C in July. In June and August, the general patterns of air temperature distribution are the same as in July with lower air temperature values (ranging from +7 to +21 °C). The air temperature difference between the north and south of the territory is greatest in May, when in the south-west the monthly air temperature increases to +16 °C and in the northern areas the air temperature is below 0 °C.

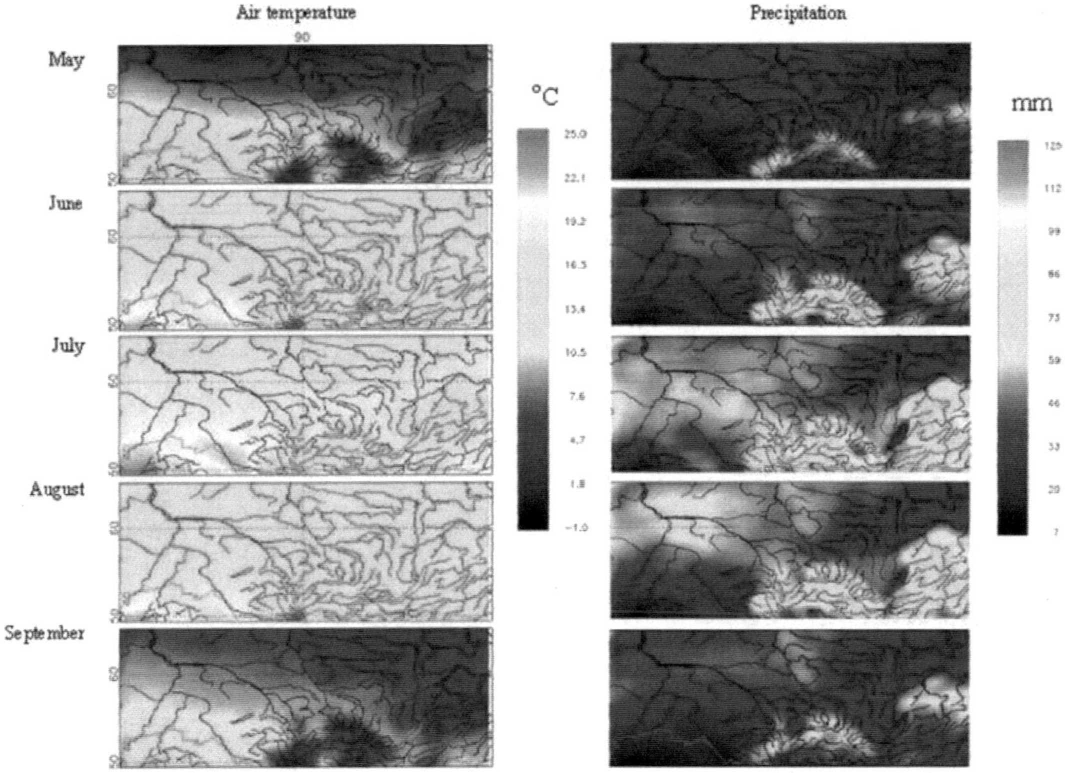

Figure.1. Long-term (1979-2010) average monthly air temperature and sum of precipitation.

The spatial distribution of trends of the monthly air temperature varies during the warm period (Figure 2). In May, for the most part of the study area, the air temperature increases with a rate of 0.3-0.7 °C / 10 years. Maximum trends (up to 1 °C / 10 years) are observed in the central part of Western Siberia. Negative trends (about –0.2 °C / 10 years) are recorded in the Altai-Sayan region and to the north of the basin of the Nizhnyaya Tunguska river. The temperature decreases in June and July in Western Siberia, but in Eastern Siberia the temperature increases. The maximum difference in the distributions of trends over the territory is observed in July in the south. The temperature trends vary from –0.65 °C / 10 years in the south-west to +1.2 °C / 10 years in the south-east. At the same time, the temperature trends in the north are minimal in modulus in July. In August and September, in the west and south-east an increase in the air temperature is observed (with rates from +0.4 to +0.8 °C / 10 years). No significant trends are observed in the rest of the territory (with rates from –0.05 to +0.2 °C / 10 years).

The distribution of monthly precipitation over the territory differs from the distribution of monthly air temperatures. The spatial distribution of precipitation for each month is similar. The maximum value (up to 125 mm) is recorded in the mountain areas (Altai, Western and Eastern Sayan, Khamar-Daban, Stanovoy Highland), and a minimum (less than 10 mm) is observed in the south-west of the territory. The greatest moisture is observed in the study area in July, and the least moisture is observed in May. The trends of monthly precipitation during the warm period are different. In most cases the trends are negative. The minimum precipitation reduction (with rates of about –30 mm / 10 years) is observed in Eastern Sayan in July. The maximum increase in precipitation (up to +14 mm / 10 years) is observed in the central part of the territory in the basins of the Ob and Yenisei rivers.

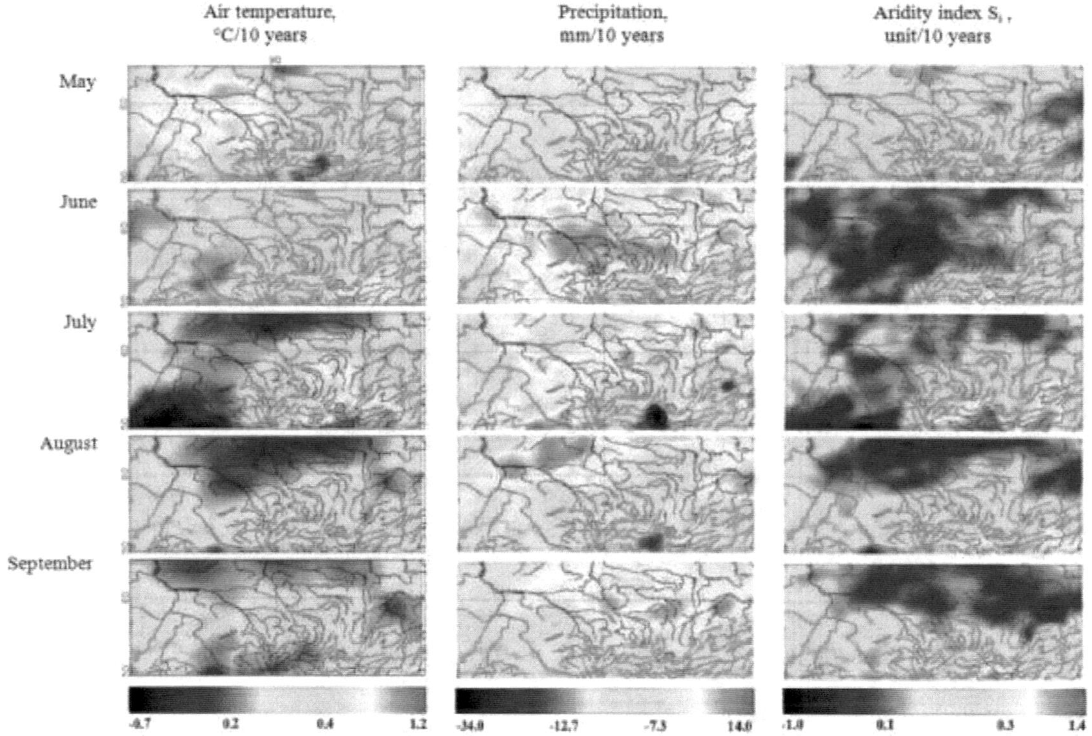

Figure.2. Trends of climate characteristics for 1979-2010.

The changes in the complex hydrothermal Ped index S_i were analyzed. The hydrothermal conditions in different months of the study area are not the same. In May, for most of the territory the aridity increases. A small decrease in S_i is observed only in the mountains of Transbaikalia. In June, the moisture increases in the central regions of Western Siberia, while the south of Eastern Siberia becomes more arid. In July, a stable increase in the S_i index occurs. The trends of the index vary from -1.0 to +1.4 units / 10 years. The area of the maximum increase in moisture shifted to the northern regions of Kazakhstan. In August, an increase in aridity is observed in the southern regions of the study area, and a decrease in aridity is observed in the northern regions. A latitudinal type of trend distribution exists in September. The trends are distributed from positive values in the south to negative trends in the north of Siberia.

4. Conclusions

According to the results of our study, the mountainous areas of Eastern Siberia are becoming more and more arid each month. In Western Siberia, aridity increases in May and decreases in June. In July, August, and September positive and negative trends are evenly distributed. The increase in aridity of the territory is associated with both the increasing temperature and decreasing precipitation. Conversely, in the areas where an increase in precipitation and a decrease in average monthly temperature are observed, some excessive moisture is observed. The greatest differences between the distributions of trends of the S_i index itself and its components are observed in July.

The results of the study are in good agreement with the conclusions obtained earlier in the analysis of droughts and periods of excessive moisture made in [16]. However, in the previous studies only point data of weather stations were analyzed. In the present paper, some spatial hydrothermal characteristics obtained from the ERA-Interim reanalysis data were studied. A good agreement between the results of the study was obtained due to the preliminary correction of the reanalysis data.

The developed software module and the web-GIS "CLIMATE " significantly facilitate and speed up the calculation of the aridity/moisture index for the study area using observation and reanalysis data. The web-GIS "CLIMATE" provided a tool to calculate and visualize the hydrothermal conditions using the Ped index for the territory of Southern Siberia from 1979 to 2010. It allowed us to estimate the spatial distributions of hydrothermal conditions for the study area. The results of the calculations are represented in net CDF files, georeferenced maps, and layers available as downloadable files via web services [17]. They can be easily transferred to online and desktop GISs for further cross-analysis with other spatial products, e.g. for the estimation of the impact of droughts on vegetation productivity. Since the hydrothermal indices are integral characteristics of the heat-moisture conditions of an area, special attention should be paid to the quality of the initial data. Adjustment of reanalysis daily precipitation data using the bilinear regression equation allows us to correctly estimate the spatial distributions of the Ped index, to increase the quality of droughts detection and estimation of excessive moisture conditions in the study area.

Acknowledgements

The work was partially supported by SB RAS (Fundamental Research Program project IX.138.2.1).

References

[1] Kattsov V M *et al* 2014 *Second Roshydromet assessment report on climate change and its consequences in Russian Federation* (Moscow: Federal service for hydrometeorology and environmental monitoring (Roshydromet)) 56

[2] Voropay N N, Maksyutova E V and Balybina A S 2011 Contemporary climatic changes in thePredbaikalie region *Environ. Res. Lett.* **6** 045209 9

[3] *Hydrometeorological disasters* 2001 ed G S Golicin and A A Vasilev (Moscow: KRUK Publisher) 296

[4] Meschcerskaya A V, Blazhevich V G, Golod M P and Belyankina I G 1986 Long-term fluctuations of aridity indexcs in warm period of year in main agricultural zone of USSR *Proc. of the Voeikov Main Geophysical Observatory* (Leningrad: Hydrometeo Pub) **505** 120-9

[5] Meschcerskaya A V, Blazhevich V G and Zhitorchuk U V 1978 Hydrothermal coefficient and connection with yields of agricultural crops *Proc. of the Voeikov Main Geophysical Observatory* (Leningrad: Hydrometeo Pub) **400** 134-49
[6] Ped D A 1975 About index of drought and excessive wetting *Proc. of the Hydrometcentre of USSR* **156** 19-38
[7] Selyanoinov G T 1928 About agricultural estimate of climate *Proc. of Agricultural meteorology* (Leningrad: Hydrometeo Pub.) **20** 165-77
[8] Rauner Yu L 1981 *Climate and productivity of cereals* (Moscow: Nauka) 163
[9] Drozdov O A 1980 *Droughts and dynamic of wetting* (Leningrad: Hydrometeo Pub) 96
[10] Pokrovskaya T V 1979 The relationship between grain crop yields in Northern Kazakhstan and meteorological conditions *Proc. GGO* (Leningrad: Hydrometeo Pub.) **403** 22-8
[11] Zolotokrylin A N, Cherenkova E A 2013 Moistening tendencies in grain-growth zone of Russia in beginning of 21st century *Problems of Ecological Monitoring and Ecosystem Modelling* **25** (Moscow: IGCE Roshydromet and RAS) 251–64
[12] Gordov E P, Okladnikov I G, Titov A G, Bogomolov V Yu, Shulgina T M and Genina E Yu 2012 Geo-information system for investigation of regional climatic changes and first results obtained *Atmospheric and Oceanic Optics* **25** 137-43
[13] Riazanova A A, Voropay N N, Okladnikov I G and Gordov E P 2016 Development of computational module of regional aridity for web-GIS "Climate" *IOP Conference Series: Earth and Environmental Science* **48** doi:10.1088/1755-1315/48/1/012032
[14] Dee D P *et al* 2011 The ERA-Interim reanalysis: configuration and performance of the data assimilation system *Quarterly Journal of the Royal Meteorological Society* **137** 553–97
[15] Shulgina T M 2012 Development and application of software complex for regional climate changes analysis using modelling data *Synopsis of the Ph.D. thesis* (Tomsk) 15
[16] Voropay N N, Maksyutova E V and Riazanova A A 2016 Hydrothermal conditions at the south of East Siberia during the ongoing warming *IOP Conference Series: Earth and Environmental Science* **48** doi:10.1088/1755-1315/48/1/012003
[17] Okladnikov I G, Gordov E P, Titov A G and Shulgina T M 2015 Information-computational system for online analysis of georeferenced climatological data *Proc. of the XVII Int. Conf. on Data Analytics and Management in Data Intensive Domains(October 13–16, 2015, Obninsk, Russia)* ed L Kalinichenko and S Starkov online CEUR-WS.org/Vol-1536/paper11.pdf 76-80